CLOUDS ARE NOT SPHERES

A Portrait of Benoît Mandelbrot

The Founding Father of Fractal Geometry

CLOUDS ARE NOT SPHERES

A Portrait of Benoît Mandelbrot

The Founding Father of Fractal Geometry

Nigel Lesmoir-Gordon

BECTU, The Media and Entertainment Union, UK

World Scientific

NEW JERSEY • LONDON • SINGAPORE • BEIJING • SHANGHAI • HONG KONG • TAIPEI • CHENNAI • TOKYO

Published by

World Scientific Publishing Europe Ltd.
57 Shelton Street, Covent Garden, London WC2H 9HE
Head office: 5 Toh Tuck Link, Singapore 596224
USA office: 27 Warren Street, Suite 401-402, Hackensack, NJ 07601

British Library Cataloguing-in-Publication Data
A catalogue record for this book is available from the British Library.

CLOUDS ARE NOT SPHERES
A Portrait of Benoît Mandelbrot, The Founding Father of Fractal Geometry

ISBN 978-1-78634-474-8

For any available supplementary material, please visit
http://www.worldscientific.com/worldscibooks/10.1142/Q0138#t=suppl

Designer: Jimmy Low

FOREWORD

Ian Stewart

Emeritus Professor of Mathematics, Warwick University

I first heard about fractals and Benoît Mandelbrot — the man whose name will forever be indelibly linked to them — from Martin Gardner's Mathematical Games column in *Scientific American*. It was typical of Gardner to present new advanced ideas in mathematical research in his friendly, informative column, and it was typical of Mandelbrot to have stimulated the growth of an area of mathematics that could be presented that way.

The prehistory of fractals goes back to the late nineteenth century, when a small coterie of mathematicians became interested in 'pathological' sets — curves that fill an entire square; curves that are infinitely long but enclose a finite area; curves that cross themselves at every point. But these fascinating shapes languished for decades, confined to a specialist area of pure mathematics. It was Mandelbrot who brought them out into the light, gave them a memorable name, and showed us how to use them to understand the natural world. There are more fractals in nature than there are circles, but mathematicians had devoted far more effort to circles than to fractals. Mandelbrot changed that.

Every new area of mathematics needs three types of persons to make it flourish: the visionary, the technician, and the publicist. The visionary sees further than others, appreciates the promise of new ideas when everyone else sees little more than a faint haze, and recognises the big picture. The technician carries out the rigorous analysis, formalises the concepts, proves the theorems, and establishes the intellectual framework. And the publicist draws the world's attention to what emerges. Mandelbrot was two of these: the visionary and the publicist. He was mathematically talented, but he was not a conventional academic: he spent most of his working life at IBM's Thomas J. Watson Research Center in New York State. He was not a typical industrial researcher either. In fact, he was not a typical *anything*. He was unique.

Ten years ago my wife and I had breakfast with him in Santa Fe, at a conference on financial mathematics. Many people found him prickly — he gave short shrift to anyone who lacked the imagination to understand why fractals were important — but on that morning he was urbane and charming. He was bubbling over with excitement because something that he had conjectured many years earlier had finally been proven. It was about Brownian motion, the irregular jiggling movement of tiny particles in a fluid, caused by molecules bumping into them. Mandelbrot had guessed an answer on the basis of computer-generated images. Now the mathematicians had confirmed his guess. What really impressed me was his joy at the solution. He did not use it to show how clever he had been. He was just enormously pleased that someone had completed the difficult task of proving that his guess really was correct.

Benoît Mandelbrot belongs to a lineage of mathematical mavericks which can be traced back to Pythagoras, Johannes Kepler, and Isaac Newton — visionaries who sought and found simple harmonious patterns concealed within the apparent complexities of nature. His work has been so influential that for many scientists, in a wide range of fields, fractals are now just another routine part of the intellectual toolkit: one of the highest honours that science can bestow. His mathematical legacy has enriched our world beyond measure. With its stunning pictures and lucid text, *Clouds Are Not Spheres* celebrates the works and the life of this remarkable man.

CONTENTS

Clouds are not Spheres
Mountains are not cones
Coastlines are not circles
Bark is not smooth
Nor does lightning travel in a straight line

Benoît Mandelbrot

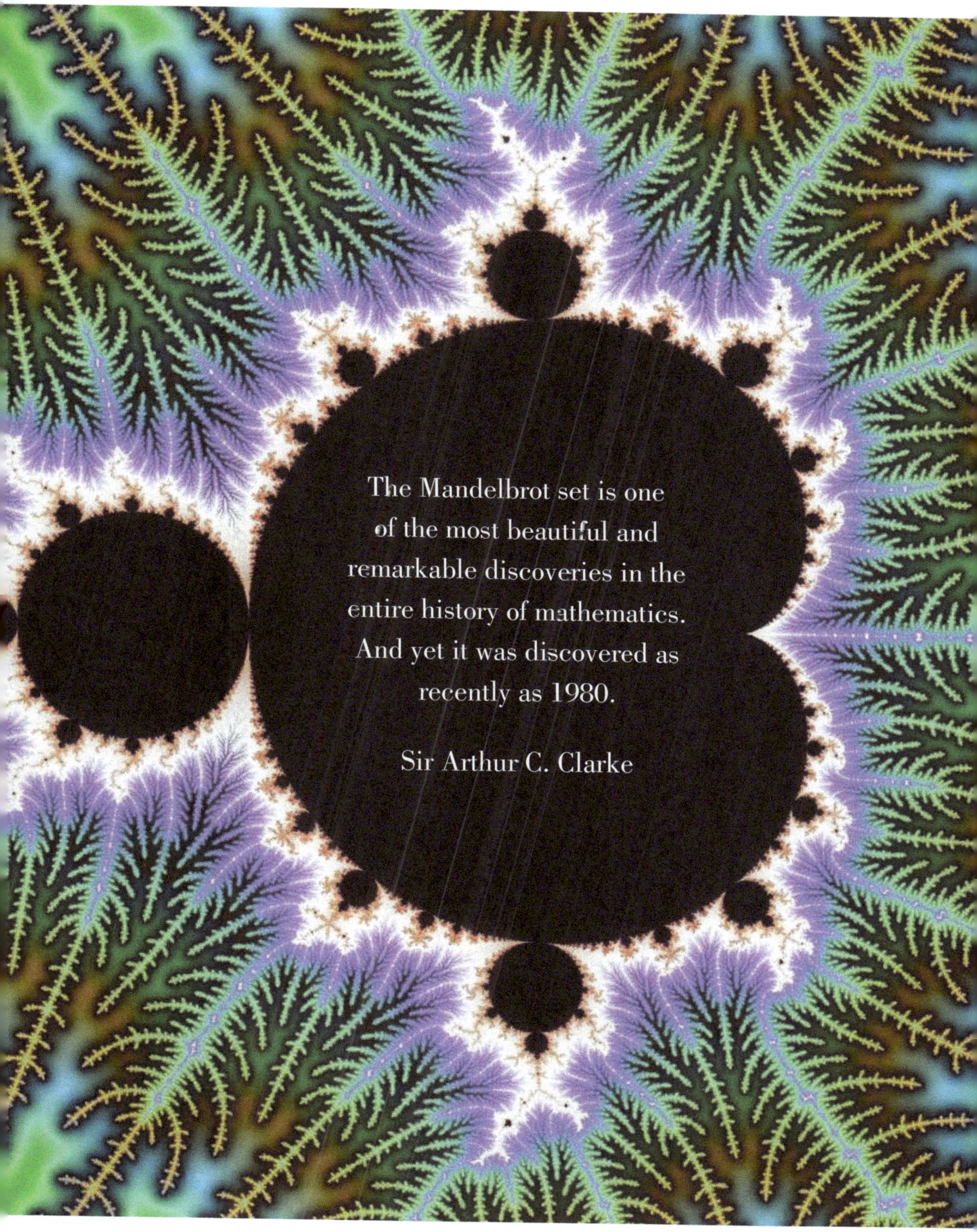

The Mandelbrot set is one of the most beautiful and remarkable discoveries in the entire history of mathematics. And yet it was discovered as recently as 1980.

Sir Arthur C. Clarke

PREFACE

This book is largely based on interviews I recorded for my film called *Clouds are not Spheres — The Life and Work of a Maverick Mathematician*, supported by some additional material from my films *The Colours of Infinity* and *Mandelbrot's World of Fractals*.

During the course of the filming for Clouds are not Spheres, I spent two days in Paris with Benoît Mandelbrot in an artist's studio where he told me his life story. Benoît was eloquent and enthusiastic, never losing his way. This interview formed the back-bone of the production. I completed this film with interviews from Nobel Laureate Professor Ivar Giaever; Yale University's Mathematics Professors Peter Jones, Michael Frame and Ronald Coifman; Warwick University's Emeritus Mathematics Professor Ian Stewart, Michael Barnsley, Ralph Gomory, and Heinz-Otto Peitgen; physicists Carl Everstz, Martin Gutzwiller and Bernard Sapoval; the science fiction author Sir Arthur C. Clarke; the philosopher Pierre Roubinet; the composer Charles Wuorinen; and the economist Richard Olsen.

INTRODUCTION

Professor Benoît Mandelbrot was Sterling Professor Emeritus of Mathematical Sciences at Yale University and IBM Fellow Emeritus (Physics) at the IBM Research Center. He was also a Fellow of the American Academy of Arts and Sciences, a Member of the U.S. National Academy of Sciences and the American Philosophical Society and a Foreign Member of the Norwegian Academy of Science and Letters. He died at the age of 83 on 20th October 2010. He felt that his work was incomplete.

I want to convey a feeling I have about the mathematical beauty of nature, about how she works inside. To realise that the phenomena we see result from the complexity of the inner workings between atoms is dramatic and wonderful. So, it's a feeling of awe, this feeling I have about the glories of the universe.

Richard Feynman

Physicist & Nobel Prize Winner

FRACTAL GEOMETRY

The world that we live in is not naturally smooth-edged. The real world has been fashioned with rough edges. It's a wiggly world! Smooth surfaces are the exception in nature.

As Alan Watts wrote: *The formless world is the Wiggly World — just like a cloud. The formal world is the world human beings are always trying to construct all the time out of rectangles and squares. We forget that we are all wiggly.*

Throughout recent human history mankind has accepted a geometry that only describes shapes rarely — if ever — found in the real world. The geometry of Euclid describes ideal shapes — the sphere, the circle, the cube, the square. Now these shapes do occur in our lives, but they are man-made and not nature-made. Uniform, rectangular objects like boxes and buildings do not occur in nature.

Benoît Mandelbrot has given science a new language to describe roughness and the way things really feel, look and sound. Fractal Geometry is an entirely new way to study and describe the real world.

Michael Frame: "The discipline has opened up a host of new directions in science. It is a marvellous way to introduce people to science. It is not just about the extremely large like galaxies or the extremely small like atoms: it happens right where we live and this language can be taught in an hour."

The invention of the silicon chip in the 1970s created a revolution in computers and communication and transformed our way of life. We are now seeing another

revolution which is changing our view of the Universe and giving us a far better understanding of its working. That revolution is in fractal geometry and the discovery of the Mandelbrot set. Fractal geometry is a deeply powerful extension of classical Euclidean Geometry.

Two distinct strands of thought converged to form Benoît Mandelbrot's fractal geometry. There was scattered evidence from various areas of experimental science which had been ignored by the scientific establishment on the grounds that it did not fit in with existing theories. Meanwhile mathematicians were trying to understand the true nature of space, relying purely upon logic and intuition. To push their theories to the limits they constructed the strangest shapes they could conceive. Some called their creations monsters. Some even looked like dragons. It took a long time for anyone to realise that in this bestiary of weird forms are found all the varied and intricate patterns of nature.

Had we never seen the stars, the sun, and the heavens, none of the words which we have spoken about the universe would ever have been uttered. But now the sight of day and night, and the months and the revolutions of the years, have created numbers and have given us a conception of time, and the power to enquire about the nature of the universe.

Plato

Three centuries before Christ, Euclid laid down the rules which were to define the subject of geometry for a millennia to come. The shapes that Euclid studied — straight lines and circles — proved so successful in explaining the universe that scientists became blind to their limitations, denouncing patterns that did not fit in Euclid's scheme as *counter-intuitive* and even *pathological*.

A steady undercurrent of ideas starting in the nineteenth century with discoveries by Karl Weierstrass, George Cantor, Felix Klein, Henri Poincaré, Carl Friedrich Gauss, Nikolai Ivanovich Lobachevski and János Bolyai led inexorably towards the creation of a whole new kind of geometry, with the power to describe aspects of the world inexpressible in the basic language of Euclid.

Michael Barnsley: "Partly because Euclid's geometry was considered to be the only geometry, no one bothered to question or examine many of its details. Towards the end of the nineteenth century, another view emerged and Euclidean geometry is now regarded as one example of many possible geometries."

Newton tamed the complex motions of projectiles and planets with the notion of differentiation. Differentiation gives the rate of change of a variable. Rate of change is the key. Systems controlled by physical forces like gravity can be analysed in terms of this rate of change. Combining the rate of change at each instant in time defines the evolution of a system. This is one of the most powerful tools in mathematics. Like evolution itself and many other scientific discoveries, the calculus was conceived simultaneously by two different men — Isaac Newton and Gottfried Wilhelm Leibniz. Newton used the calculus to develop his celebrated theory of universal gravitation, whereas Leibniz developed the clearest formulation of the calculus. The Leibniz notation survives to this day.

Maxwell's equations for electromagnetism and Clausius' laws of thermodynamics flowed from this empowering discovery. Virtually every branch of science was transformed. It was assumed that all phenomena could be analysed in terms of this new understanding. In 1850, Pierre Laplace claimed that given the position of every particle in the universe and its rate of change, we could predict the entire future of the universe forever in its every detail.

Mandelbrot admired Bernard Bolzano's work and thought that his connection to fractal functions was not as well known as it should be. The first example of a mathematical fractal was made by Bolzano in 1830. He proved that arbitrarily close to every point of the graph are infinitely many points at which the function is non-differentiable. The proof that the function is nowhere differentiable came later.

In 1861, Karl Theodor Wilhelm Weierstrass's quest for absolute rigour led to his discovery of a continuous function that is nowhere differentiable. It was just not possible to define its rates of change. This was a shock to the scientists of the day. That a curve should exist with these properties was widely deemed to be highly counter-intuitive. It was intuition that would have to change. Weierstrass wrote that a mathematician who is not also something of a poet will never be a perfect mathematician!

Weierstrass put infinitesimals on a firm footing by defining them in terms of limits of infinite sequences. His new methods opened up a whole new branch of mathematics called Analysis. Analysis arose as an attempt to bring a new rigour to mathematics. Precise notions of number and continuity were sought. All of mathematics could be explained in terms of simple whole numbers. But where these came from was a mystery.

On 1 March 1980 at IBM's Thomas J. Watson Research Center in upstate New York Benoît Mandelbrot discovered the now iconic Mandelbrot set. With its thrilling visualisations and infinite nature, it brought the world of mathematics back into public consciousness.

Arthur C. Clarke has called the set one of his seven wonders of the world. In the film *The Colours of Infinity*, which Arthur presents, he says: "*No matter how much we magnified it, a million times, a billion times — until the original set was bigger than the entire universe — we would still see new patterns, new images emerging, because the frontier of the Mandelbrot set is infinitely complex. And when I say infinitely, I really mean that. Most people when they say infinitely, they mean 'Oh, only, rather big.' But, this is really infinity!*"

Gaston Julia

The seeds of this discovery were in fact sown decades before the M-set was first seen. In Paris, in 1917, two mathematicians, Gaston Julia, a student of Henri Poincaré, and Pierre Fatou, published papers connected with complex numbers. The results of their endeavours eventually became known as Julia sets, although Julia and Fatou never saw a Julia set! They could only guess at them. And it would not be until the advent of modern computers that Julia sets could be seen for the first time. It was Mandelbrot's uncle Szolem, who initially directed him to the work of Gaston Julia and Pierre Fatou on self-similarity and iterated functions.

Fractals are aesthetically pleasing, frequently revealing stunning beauty in the most subtle ways. The Mandelbrot set defies verbal description. Fractals seem intimately connected to the concepts of beauty and elegance.

Amazingly, until recently, there was no word to describe the familiar shapes of nature. Now we can see that there are fractals everywhere. Now we can view the universe through fractal eyes.

Fractal Geometry will make you see everything differently. You risk losing your childhood vision of things — of clouds, forests, galaxies, leaves, feathers, flowers, rocks, mountains, torrents of water, carpets, bricks, people, dogs, and much else besides! Never again will your interpretation of these things be quite the same again.

Michael Barnsley

01

THE EARLY YEARS

Benoît Mandelbrot was born in Warsaw on 20 November 1924. His parents were Lithuanian Jews. They met in Warsaw, married and had one child before Benoît. Their first child died, but Benoît did eventually have a younger brother. Although his parents were *unusual*, he recalled "I had a very sheltered and happy early childhood in Poland."

Benoît's father, Charles, was a clothing manufacturer and retailer. He was self-taught and not formally educated. Family circumstances contrived that Charles was also responsible for his brother Szolem's education and upbringing. Importantly, Szolem Mandelbrot would eventually become a highly successful mathematician in France and very much Benoît's 'mathematical uncle'.

'On my father's side', Benoît remembered, "they were all scholars as far as anyone can remember — for centuries and centuries. My uncle Szolem, the younger brother of my father, became a very well-known mathematician in Paris."

His mother had lost a child in an epidemic and did not want Benoît to die in another one. It was very dangerous in terms of epidemics and other complications and his mother kept him out of school for several years. He had another uncle, Loterman, who tutored him. Benoît describes him as an amateur, who only taught him to play chess and told him stories from antiquity. He advised Benoît to learn the alphabet and the multiplication tables but he never did.

Benoît and his family in Warsaw

Benoît did however show an early delight in geometry. He excelled at chess, though he admits that he did not think the game through logically, but in a geometric fashion. Maps were another important inspiration in the development of his extraordinary geometric abilities. His father was crazy about maps and there were maps all over their house. Benoît read voraciously and widely.

He recalls, "I didn't go to the first two years of elementary school, then coasted later on. I came to Paris and started two years late because of the change of the school system. I was not

Benoît and his younger brother

© Aliette Mandelbrot

really present in class. I was dreaming and so on without any particular difficulty because I was reading a lot on my own. In general, I had a life which was not very much school-bound."

In 1936, when Benoît was 12, his family had the foresight to quit Poland. Nazi persecution loomed dark and heavy over them. "My mother gave up her profession as a doctor and became a homemaker. My father was bankrupted. It was a very difficult time. We were moving away from friends and family to a new and unknown country. We came to Paris. I had an aunt there, my father's sister. She was the only person we knew. She certainly saved us from death and extreme depression."

Benoît's uncle Szolem, who had learned French through studying the work of Henry Poincaré, became a professor at the University of Clermont-Ferrand when Benoît was five. When Benoît turned 13, Szolem was appointed a professor of mathematics at the College de France in Paris proving that a career in this subject was a real possibility. Despite Szolem's success, his father still encouraged Benoît to study something practical. But his uncle's appointment made a deep impression on him.

Benoît started at the Lyceé in 1937. He was already two years older than most of his peers. Looking back he recalled, "Before the war France had a highly organised school system. Students went to higher primary school until they were 18. Elementary school students graduated with an excellent education. At that time in France when students were as young as six, they began preparatory classes for high school, which started at around age 11. So a child's future was decided between the ages of 11 and 13.

Being a year behind was a formidable obstacle. It was because my French was shaky. The written and oral examinations were tough. The certificate was the big thing. One thing I was good at was spelling because I could visualise things. The written exams were therefore easier for me. I had a good time actually in the first two years of high school. I don't remember much, but I did enjoy it."

When Paris fell to the Nazis in 1940, the family fled further south to a little town in central France called Tulle. "My uncle had a country house nearby which is why we went to Tulle. Tulle, which is in the hills in central France, is a very poor, arid part of the country with deep valleys and low plateaux."

Benoît started jumping years in school. "I had an easy time in Tulle. It was in fact ridiculously easy for me there. I was a high-pressure kid from the big city. At this time I was more interested in literature and history than in mathematics. Maths was easy for me, but it was not my passion." At one point, Benoît became an apprentice tool-maker. He recalled: "I did this work surprisingly well because I had a firm hand and a very good feeling for the shape of things."

He made many friends in Tulle. There was one friend in particular who eventually became a significant and long-standing ally, the philosopher Pierre Roubinet. They met on 1 October 1939. Roubinet recalls that when they started their last year at Tulle's Secondary School during the first break, they spoke for the first time. "Or more precisely Benoît spoke and I listened and responded to him. Then at midday too, and so on and on for 60 years! Benoît enjoyed talking because he enjoyed communicating. He liked to mentally construct a universe of forms. He enjoyed enquiring, building forms within himself — particularly the forms of nature and the countryside. He loved painting." Roubinet remembers that Benoît liked describing the architecture and the planning of towns and cities like Barcelona, Berlin and Paris. Benoît was very descriptive. Roubinet says that, "With regard to his schooling, it's really a poem. Benoît was always an outstanding student, outstanding in Tulle's Lycee, with Mention Tres Bien in Elementary Mathematics, Mention Tres Bien in the Baccalaureat of Elementary Mathematics and other distinctions. But Benoît was never completely integrated, at least not in his youth, and even later on in his life in the school system he always had difficulties."

In Tulle's Lyceé, people had a high opinion of him because he was a true gentleman and also an outstanding pupil. He enjoyed talking to everybody. He was bright and courteous, but he was an outsider. Roubinet mentions again: "Benoît evidently made his own education, an education intellectually far brighter than the one given to the little Tulle countrymen which we were! Also Benoît was working with his uncle Szolem, himself a mathematician, who was already a member of the

Academie des Sciences."

Benoît recalled: "Then things became very bad. The Germans occupied Southern France which they had not in 1940. And I spent a year and a half running around, keeping body and soul together." His education was haphazard during this period of Nazi occupation. As Benoît put it, "If I had been caught at that time and fallen into Nazi hands, I would have died; but that's a long story. The important thing was that I did not go to school for a while. However, when I did eventually get back to school, my talent for geometry saved me."

When Heinz-Otto Peitgen looks back on Benoît's early life, he says: "His childhood was certainly not an easy one, although when he talked about it he always asserted that it was a very rich, very positive one. But things were rough. When he came to France in 1936, things were heading towards the terrible period of the German occupation and Benoît had to go through a lot. But in those years in southern France he always picked up something fruitful, something positive, something which took him forward another step. That was his nature. No matter where he was, no matter how things were seen by other people as rough or not productive or unfortunate, Benoît always made the best of it. He picked up whatever he could and he put everything together in a very positive way."

Roubinet: "I must say that being a philosopher, I can't help making the comment that Benoît was like Aristotle. He had the same concept of knowledge and of transmitting knowledge. For Aristotle, knowledge consisted of constructing images. Benoît liked to construct shapes too that were very often visual in the same way as Aristotle. He had the same theory on information and the distribution of information. Benoît felt an immense pleasure in communication and I also took immense pleasure in listening to him. He was a great source of joy. He was an enthusiastic man who truly communicated."

Benoît went to Lyon for a year of post-High School study in mathematics. It was here that he first discovered that he had a remarkable and quite extraordinary visual ability. The major high schools in France had special post-graduate years: some in mathematics, some in the humanities, some in other practical subjects. He went into mathematics. He recalls that the first week was a total blur. He did not know what was happening. He was expected to "just keep quiet and survive", he said. The second week was also a blur to him, but in the third week something extraordinary happened. He discovered he had a gift which he had been totally unaware of. He could transform any problem spontaneously which the teacher had set up to be a very difficult algebraic problem. He could transform the problem instantly into shapes. When the teacher asked for a solution to an equation Benoît would see two shapes intersecting or two shapes projecting into each other.

Once he saw them as geometric problems, the answer was immediate. He would lift his hand and boldly declare, "Well, Sir, what you're asking is an intersection of this shape with that shape." The professor would respond with, "Are you sure?" Then he would check himself and find that Benoît was absolutely right. As the term progressed in that winter in Lyon in 1944, he became more and more aware of the strength and the reliability of his freakish gift. He could invariably take the most abstract problems, which the teacher posed and, without any effort, without any conscious searching, see them differently.

He recalls, "This revelation of January 1944 revealed the principal tool — my principal strength. It allowed me to go into many different fields and to transform problems in each one of them into geometric shapes — or a collection of shapes — which I could live with, understand, become acquainted with, describe first intuitively and then mathematically."

Benoît was always the first to concede that his uncle Szolem had been a most important influence in his life. But his uncle was very careful to keep his mathematics on working days completely dry and with no pictures! On the weekends however he was a 'Sunday painter'. In due course, the shapes, the 'Sunday painting' and the mathematics came together. Benoît was able to integrate them completely. His lack of schooling was very much a part of his unique development. He skipped many grades in elementary school, then some grades in high school and even some grades at post-graduate school. This was a perfect preparation for when he would eventually have to jump 'into the cold'. Had he been in a regular, well-organised educational system not perturbed by war and by immigration to France that gift might have been smothered and have never revealed itself.

Looking back at those days, the physicist Martin Gutzwiller appreciated that Benoît showed a tremendous survival instinct like so many, who went

Benoît the schoolboy with a friend

through the terrible years of Nazi persecution in France. "The ones who survived showed special qualities. They could take care of themselves. Benoît found himself in many tight spots where he could have been caught, but he always got himself out of them."

After the liberation of France at the end of the war, Benoît returned to Paris and cramming school to prepare for examinations. During his third week in class, he bravely told his professor that the methods he was using were a waste of time. Benoît told him that he needed only to imagine the shapes that the equations would give and then the answer would be obvious. Extraordinarily, Benoît ended up passing the entrance exams without the usual two years of preparation! Most students took two years, three years to prepare, even four in some cases. But still they failed. Thanks to his grasp of geometry, he was able to solve problems which other people found almost impossible. He was so ahead in mathematics that his terrible grades in other fields were found acceptable. However, he did do very well in French, in English and in drawing. By a fluke, one of the examinations included drawing because it was an engineering school. Traditionally, engineers were supposed to be able at least to draw an outline shape, so drawing was absolutely essential in this discipline. As it happened, most of the other students were poor in this particular area. Benoît described himself as being 'dull but very accurate'. His drawings were not interesting, but they were extremely precise.

The examinations for the great schools in Paris should have been held in June of 1944. However, because of the war, they were finally held in December. They were indeed 'ferocious examinations'. They lasted a full month. By the time they were completed, it turned out that Benoît had completed the examinations for two very elite schools. He found out that he could, in a way, 'cheat quite legally'. He broke no rules, no written laws. Instead of doing very complicated calculations he could quite simply see what was happening. He really knew what was happening. He could describe things visually and did not have to search for answers. During the second period of the oral exams, his teacher asked Benoît if he could speak with him.

"Well, yes, certainly you can", Benoît replied.

The teacher asked, "Did you do the maths problem?"

"Yes," responded Benoît.

"Well," the teacher went on, "All the teachers and the examiners tell me that only one student in the whole of France solved the problem and he was in my class. I couldn't do it myself and none of the other students in my class could do it. So I thought it might be you. Was it?"

"It was me," replied Benoît confidently.

The teacher was amazed. "How did you do it? There is a truly horrible triple integral in the middle."

Benoît laughed. "It's very simple", he said. "It's the volume of a sphere. If you do it right. If you look at it right, not at the x, y, z, but at the other co-ordinates."

The teacher was genuinely shocked and went away shaking his head, muttering, "Of course, of course, of course." He appreciated immediately that it really was that simple.

Benoît started to study mathematics at the Ecole Normale Supérieure which trained university and high-school professors. But mathematics was dominated then by a group known as Bourbaki — an assumed name. Ironically, Benoît's uncle Szolem was one of the founders of the movement long before it seriously set about attempting to purify mathematics by making it more formal. Szolem had learnt French studying the work of the renowned mathematician Henri Poincaré and became a professor at the University of Clermont-Ferrand when Benoît was five. When Benoît was 13, Szolem moved to the College de France in Paris. But despite Szolem's success, Benoît's father discouraged his son from pursuing mathematics.

When Benoît first entered The Ecole Normale Supérieure, it was small and elite. It was in principle a school for abstract mathematics and physics. After just two days at that school, something very significant happened to Benoît. He looked around him and experienced a violent feeling. He asked, "What am I doing here? It's the wrong place for me."

It hit Benoît that the emphasis at that school was focused far too much on very abstract mathematics. He knew from his uncle Szolem that the school and the Bourbakists were pushing further and further in the direction of the abstract. The group wanted to go way beyond anything that had been done before in this direction. The Bourbakists were taking over.

Benoît: "They had a determined plan for this school which did not hold any promise or enjoyment at all for me because the kind of mathematics that I liked was very visual. The kind of mathematics I liked was very close to nature — hard things, rough things — and that school was practising the precepts of Plato in the strongest possible way. Plato wanted mathematics to be separate from matter and sensations, but what I liked was the intimate link between mathematics,

Right:
Benoît's uncle, Szolem Mandelbrot

Far right:
The young Benoît

matter and sensation."

So Benoît did not agree with the principles of the Ecole Normale Supérieure. He also felt, and with very good reason, that he could not bear the triumphant airs and the authoritarianism of that school. He spent a day agonising and then resigned. This was a decision which set in train the path his life was to follow from then on. He had been accepted into the most exclusive school, which would certainly have given him a guaranteed future. The alternative, he took was to go to the other school he had sat the examinations for, the Ecole Polytechnique which had a less cohesive plan. It led students into a much greater variety of very different professions. Some people became scientists while others became managers of big companies or went into the military. And that really attracted Benoît because he was not seeking somewhere where he would be strongly attached to pure theory, but somewhere which would give him freedom — enough scope, enough variety of possibilities for him to create something that would fit his exceptional skills and his extraordinary personality.

Benoît stayed at the Ecole Normale Supérieure for just three days. Pierre Roubinet's room was close to Benoît's. On the evening of his second day, he went to Roubinet's room and told him that he was leaving Normale. Roubinet remembers that he found that decision 'completely barmy' and he laughed. But the Polytechnique was still a prestigious school so, in the end, Roubinet understood why Benoît was opting to go there when it was explained to him. "At the Ecole Polytechnique they left him alone to think as he pleased. This was certainly a bold decision, but Benoît was a bold man."

He realised that, "What I wanted to do", as he explained to his uncle and others, "Would be seen as totally

absurd because it was so romantic and childish. I wanted to experience the excitement of being the first in a field, to bring back the kind of rational thought which physicists had brought to physics several centuries before."

His uncle Szolem made fun of him. He joked, "Well, it was all done by Kepler and Galileo, so what is there to be done?" His uncle's influence, though, was important and positive, but also sometimes harsh and negative. Szolem had very fixed views and was highly intolerant of those who disagreed. Benoît recalled that they had many terrible rows.

02

CALTECH AND TO FRANCE ONCE MORE

While Benoît was studying at the Ecole Polytechnique, the man who had the most influence on him was a professor of mathematics called Paul Levy. This was so even though Benoît was not his student.

Paul Levy was viewed by many as being a very brilliant man, but he was also an 'odd-ball' and an eccentric of the first magnitude. He had been very badly treated by the French University because of the way he worked with probability theory. It was too intuitive for some and too strange for others. He was very much a loner.

"During my last term at Polytechnique, I kept looking for ways to use my mathematical gifts and growing knowledge in dealing with real concrete problems in nature. My hopes were thoroughly romantic: to be the first to find order where everyone else had seen only mess. Someone who heard me say this commented that my dream was to be Johannes Kepler, but that the Kepler days were over!"

The physicist Carl Everstz recalls: "Based on my direct experience working with Benoît for three years, my feeling is that his knowledge of probability theory was crucial in the development of his ideas on self-similarity and fractal geometry. I think Benoît had a very deep understanding of the work of Paul Levy. He always said that Paul Levy was not his mentor, but he certainly did have access to Paul Levy's work. And he had a very deep understanding of that work, a working understanding. With this understand-

Paul Levy

ing, he went into the various sciences. He attacked a variety of problems and slowly but surely he started to understand how applicable his sort of mathematics was going to be."

While he was studying at the Ecole Polytechnique, Benoît asked his teacher if he would recommend somewhere where he could do what he wanted to do. His teacher suggested that that might be Caltech — the California Institute of Technology — as one of the professors at Caltech was Theodore von Kármán whom Benoît referred to as 'the magician'! Von Kármán knew how to apply deep mathematical ideas in very complicated areas like aeronautics. In fact, by the time Benoît arrived at Caltech, von Kármán had already left.

Benoît spent two years there where he was, as he put it, "In many ways just biding my time". In some ways it was a very important time because he witnessed an extraordinary event: the birth of molecular biology. Benoît had enormous respect for Max Delbrück. He describes Delbrück as a 'very brave man; a very independent person'. Delbrück established a new kind of biology, which had no name then, but eventually came to be known as molecular biology. Benoît saw this happening and it impressed him deeply. He was inspired and encouraged by Delbrück because he saw that he was 'doing something, which did not yet exist. He was not working in a field which existed. He created a new field'.

Benoît moves to America

© Aliette Mandelbrot

While at Caltech, Benoît studied aeronautics where he encountered the daunting complexity of turbulence, the wildly tangled fluid motions that above a critical speed replaced stream-lined flow. This was a line of study he would come back to later. He received his Master's Degree in Aeronautics in 1948.

After Caltech, Benoît went back to France. He was a disappointed man. "He says because the miracle of finding something special to do in America had not happened." He went to work for the French Air Force where he spent a year. When asked what he accomplished there he responded with, "I studied a great deal."

He then moved to the University of Paris to look for a PhD subject which was not an easy task for Benoît! When he was visiting his uncle Szolem, Benoît asked him for something to read in the metro. Szolem reached into his wastepaper basket and pulled out a document, saying, "Somebody sent me this paper which is crazy, but you like crazy things. So here it is."

Benoît took that paper on the metro with him. It was the review of a book. The reviewer stated that the book described some remarkable property, which is true of every language — be it French, English or even Latin. This was Zipf's Law. This paper baffled everybody who looked at it. And that really stimulated Benoît. He knew that he had to find an explanation for the property described in the book. It took Benoît exactly the length of that metro ride to absorb and understand the theory. He decided there and then that his thesis was to be to a large extent on linguistics and thermodynamics.

Zipf's Law's law was an attempt to bring order to word statistics. It held that given some corpus of natural language statements, the frequency of any word is inversely proportional to its rank in the frequency table. Thus the most frequent word will occur approximately twice as often as the second most frequent word, three times as often as the third most frequent word etc. His thesis, which he completed in 1952, was called *Games of Communication*. The first part of the paper was based on his ideas about Zipf's Law and the second part was surprisingly on thermodynamics. He recalls: "The mathematical linguistics in the first was a very exotic form of the statistical thermodynamics in the second half."

Benoît finished his thesis and became a Lecturer at the University of Lille. The future outcome of his career in mathematics looked certain. He would end up as a Professor in the Sorbonne and would most certainly be a Member of the Academie des Sciences just as his uncle Szolem had been. Pierre Roubinet met up with Benoît for coffee one morning in Lille. He remembers this meeting well. Benoît told Roubinet something which really amazed his friend. He informed Roubinet: "I am leaving the University to work in the USA." Roubinet wondered if Benoît had really lost all his common sense. "Leaving the university where he had titles, a royal path to greater academic posts, seemed just amazing to me. To leave French public office where he had the certainty of a comfortable pension at the end of his career to go to America! I just couldn't believe it."

03

THE USA, FRANCE AND IBM

After leaving Lille University, Benoît went into industry. He was hired by Philips to work on colour television. This work required very concrete thinking about spectral analysis. He had an intuitive understanding of waves so he did well. After Philips, he went to the Massachusetts Institute of Technology to work on information theory. In 1952 after about six months at MIT he decided to move on to Princeton's Institute for Advanced Study where he was sponsored by John von Neumann.

John von Neumann

Von Neumann introduced Benoît to Henry McKean Jr. who introduced him in turn to the idea of the Hausdorff–Besicovitch dimension — the revelation that there were phenomena that existed outside one-dimensional space, but in somewhat less than two dimensions. Benoît adopted the Hausdorff–Besicovitch dimension on the spot. It was an almost ubiquitous tool and a special example of the wider concept of fractal dimension which was to come.

Benoît at Princeton

© Aliette Mandelbrot

Benoît recalled: "After leaving Lille University, I went back to America, to Princeton University, for a post-doc. I went to work with Von Neumann, who was best known at the time as being the man who was creating the first very large, successful computer. But I didn't go to Von Neumann just for the computer, but in general because Von Neumann was to me an example that *something was possible*. Levy was a mathematician, but one who was very much outside the mainstream of mathematics." "I felt it was nice and warming to know a man like that. He was very kind to me but he was also very busy. He left Princeton to join the Atomic Energy Commission and died shortly afterwards. I was his last post-doctoral student. I learned much later that Von Neumann had been very worried about me and that he had asked the president of a major foundation to keep an eye on me because he thought I might get into some very, very rough and tough spots. I was taking gambles which he thought people should not be taking."

Von Neumann had the reputation of being a man of all trades *and* a master of them all. He was a well-established and respected mathematician. He had done significant and fundamental work in logic. He had been deeply involved in the foundations of quantum mechanics and had produced a book called *The Theory of Games* which formed the basic foundation of a new way of looking at economics. He was also an engineer. He was designing his computer with a team of people, but he was much the key man. This was Benoît's first introduction to the world of computers. His interest in these machines was immediate. From then on, his enthusiasm for this new tool

Benoît, Aliette and Laurent

continued to evolve and never waned. Once more he returned to France where he married Aliette Kagan in 1955. They moved to Geneva where Benoît planned to work with Jean Piaget because Piaget was interested in mathematical concepts related to children and language. Unfortunately, this collaboration never materialised.

He was awarded a position as a professor at the University of Lille. His academic future looked assured. But, again, he felt uncomfortable in the academic environment and in 1958 he went to IBM as a faculty visitor for the summer. It was there that his ideas bore fruit. His first son, Laurent, was born in 1962.

Usually, a professor of mathematics in France had a highly structured career path mapped out for them. Benoît felt very strongly that he was not going to get anywhere going down that established road. He felt that the type of work he wanted to do would not be respected by the people he was working with. He found them discussing him openly with comments like: "Well, he's a brilliant man, who has somehow run astray and lost himself." He did not like that. Put simply he did not feel happy in that environment.

He was not long at IBM before he realised that that place might be just

IBM's Thomas J Watson Research Center, Yorktown Heights

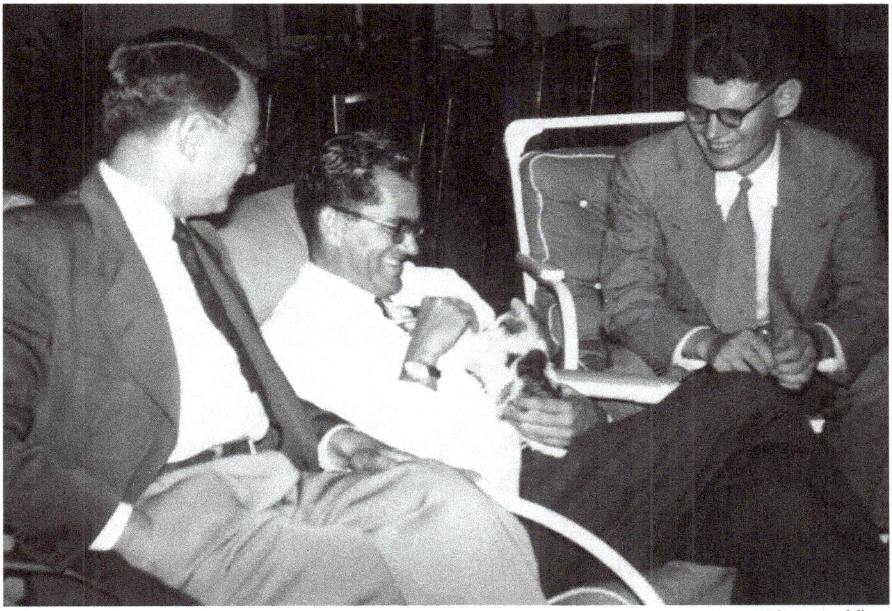

Benoît at IBM

right for him. As he put it, "That was not because it was organised to support exactly what I wanted to do, because at that stage I still had no clear idea what I wanted to do. But I knew immediately that the move was extremely fortunate and held great promise for me."

IBM gave Mandelbrot the funding, the facilities, a research team and the mental space in which to work. The powers-that-be at IBM, it must be said, had vision unlike the often reactionary management of the mainstream academic world.

Looking back, Benoît observed: "My reaction to that now is, let them live. It's not a question of enforcing uniformity in mathematics. There are different ways of doing things, different ways of finding mathematical conjectures, mathematical hypotheses. So let them live. The more the merrier!"

Reflecting on Benoît's move, Professor Martin Gutzwiller who also worked at IBM's Thomas J. Watson Center in Yorktown Heights, commented that, "It was probably better for him to be associated with IBM. It gave him more freedom and that was probably true for me too. Our personalities do not seem to be well adapted to academic life and so it worked out well for us as long as the good times lasted!"

Martin Gutzwiller

When the Yale mathematician Ronald Coifman looked back at Benoît's time at IBM he had this to say: "Being at IBM was extremely productive for Benoît on two levels. One was the nurturing environment. The other was having the computational resources at his hand before the PC was generally available and before anybody actually used the computer as a laboratory. So the combination of his view of the world, the environment, and the possibility to interact with computers was indeed very lucky for us all."

Benoît: "IBM was built up primarily by Thomas J. Watson senior to make mechanical devices like sorters. For many years this was a highly successful business. There was very little modern technology involved. It was a company based on the principles of first class manufacturing and aggressive, focused sales techniques. But that was coming to an end and Thomas Watson junior recognised that. The young Watson worked on the shell which his father had built on the skills of mechanical engineers and salesmen. He devised an entirely different company which was going to work in electronics and computers that were at this time just beginning to exist as a commercial possibility. To do this IBM required an almost total change of personnel. They engaged in a massive hiring of new people to build the revamped company which the young Watson envisaged."

The Nobel Laureate Ivar Giaever contends that IBM was very good to Benoît. "They were very open. They supported him for many years. I was working for General Electric when I got the Nobel Prize. General Electric supported me for many years. At that particular time, corporations had more faith in science than they have today and they were willing to support people like Benoît and myself. I think that IBM deserves a lot of credit for giving him the time and the freedom to pursue what he sought to achieve."

As Benoît recalled, "Many people joined and many people left IBM in a period of great turmoil. The company took on a group of people who did not feel comfortable in the academic environment of the time that were nonetheless very rich because university science departments were affluent and well-endowed. Many years later it turned out that those people taken on by IBM performed well beyond anything that might have been expected of them. They were not very well defined straight people with a clear career path ahead of them."

IBM's Dr Ralph Gomory took an immediate interest in Benoît's work and has remained a loyal supporter and friend since their first meeting.

Ralph Gomory: "I think my role was, in some sense, a helper. When I first met Benoît, I realised that he was not quite your ordinary person. I was among the very first to appreciate the depth and scope of what he was doing and I was able, because of my position at IBM, to help propose him for awards and to help him to get the recognition that was really late in coming, but which was certainly very richly deserved."

Dr Carl Evertsz: "I would say having to search for grants continuously would have been a killing factor for the sort of work Benoît was doing." "I think that his being at IBM with Ralph Gomory, who saw that Benoît was doing very, very interesting things, was absolutely vital. The fact that Gomory let him go and do these things was instrumental in Benoît being able to move really easily from one topic to another."

Benoît decided that IBM was the place where he could get a few years in hand to do his 'gambling'. He was dying to do a kind of untried work and he was always in hope of finding some subject where a full understanding of physics would be useful.

Carl Evertsz

After the summer, he decided that going back to his position in France would be 'plainly ridiculous'. He saw himself heading into a dead end. He decided to stay at IBM because it simply suited him better.

Benoît: "The gamble was very successful and paid off because within just three years I had made my first truly major discovery. That discovery came in a field about which I knew absolutely nothing, which was then called economics and later, more precisely, finance."

Benoît the teacher

How did he arrive at this discovery? It was through a picture. Benoît was visiting Harvard to give a seminar in economics, on an aspect of his work which he said was, "OK, but not first rate".

His host had a drawing on the blackboard and Benoît asked him, "How come you have the same drawing as I will be using in my lecture?" His host replied, "I've no idea what your lecture is going to be about. This drawing is about the behaviour of cotton prices." They chatted after Benoît's lecture. Benoît decided on the spot that cotton prices deserved extremely careful attention.

He left Harvard to go back to IBM with seven boxes of cards which represented to him 'a fortune in data'. At that time, data like this was very difficult to get hold of. When he got back to IBM he made a number of tests and those tests were phenomenally successful. In a very short time, he had a model for the variations of some speculative prices. It was innovative, even daring. Though, Benoît claimed, "Once you understood the mathematics it was really a very simple idea."

Economics, like most of the younger sciences, was attempting to imitate physics because there were no alternatives at that time. But imitating physics can take many different forms. The easiest way is to imitate certain details of physics. In physics, for example, the motion of a body must be continuous. It can't just go from one place and jump to somewhere else. There is inertia. Inertia makes motion continuous. However, in truth, economics can be very jagged and irregular and not at all continuous. In economics, Benoît argued, "There's no reason why prices should be continuous. If a piece of news arrives, a share price can go from, say, $100 to $10 or to $3 instantly or to $300 instantly. If a certain company has a drug — a drug that is refused, say, by the Federal Drug Administration — it's worth nothing and that company has no value. If that drug is then accepted by the FDA the company will suddenly and abruptly have enormous value. The $100 value is an average between terrible prospects and great prospects and the likelihood is that the company's share price will go from one value to the other discontinuously." Benoît accepted the notion of discontinuity in his model of prices, but strangely that idea simply did not occur to anyone else.

After Carl Evertsz left Yale, he went to Bremen and the Institut für Dynamische Systeme. The groundbreaking work Evertsz did with Benoît and the work he did in fractal geometry over the following decade very much shaped the work he went on to do in subsequent years. Evertsz recalls that: "It was because of the work I did with Benoît at Yale that I got an interest in working on the economic time series. We never worked together on the economic time series, but it was clear from his way of thinking and from the discussions we had that the economic time series — the price records of stocks —

was a very important subject to study. This is the topic that Benoît worked on in 1960. That was a very, very influential work and a very good work. It's incredible that after 30 years when I went and read that paper again to see again how much was there already. Although the word 'fractal' was not there, everything else was described in there. For example, where he pointed out very beautifully that if you look at the stock record, whether it is a monthly record, a weekly record, or a daily record, you're not able to see the difference. And not only did he mention that it is self-similar, but he went all the way to even make models for that self-similarity and explain where that self-similarity came from. I would say that this work was already in the field of fractal geometry. It definitely shaped everything that I do now."

The idea that prices could be discontinuous was amazingly and stunningly not part of what economic theory was at that time. The economists believed that they could somehow just work around the issue. They blindly believed that prices did not jump. They just moved quickly and smoothly. Now we know that was a totally incorrect approximation.

The model that Benoît came up with back in 1963 was certainly very simplified. It worked well enough for cotton, but it did not work well for other speculative prices. In fact, it did have an element of predictability, which is of course a very rare situation in economics.

He started out examining in great detail, with great precision and honesty, the daily price changes over five years. It gave him a model. He made the machinery work and he found predictions for what would happen for a day, for a month and even for a century. He then compared his findings with what actually happened and he found they were extremely close to what his data had predicted. His conclusions are still proving hard for economists to absorb. There is no discontinuity and separation between the daily fluctuations people call speculative and the long-term fluctuations which they call deeply and fundamentally economical. In fact the two merge into each other seamlessly.

Right at the beginning of Benoît's career at IBM, he tackled a practical issue which directly involved and concerned his employers. Inside the company, data was being lost or corrupted when passing between computers by random noise bursts which they could not get rid of or predict.

The problem was an expensive one but the experts could not discover the source of the noise. The errors were not completely random. They tended to come in bunches. Benoît observed that the degree of bunching remained constant whether he plotted them by the second, by the hour by the day or by the month. He applied his own mathematics and came at the problem in a completely new way. It made no sense to IBM's engineers, but they had

Benoît and Aliette Mandelbrot

to accept that the mathematics did predict that it was just not possible to compute the average rate of errors over any period of time. What Benoît's calculations revealed was that the errors were fractal and self-similar over a wide range of scales. This was in fact a Cantor set. Several years after Benoit and Jay Berger published this result, Steven Smale showed that Cantor sets often arise in non-linear dynamics. Benoît was beginning to grasp, and explain mathematically, the way things work in the real world. He sensed the underlying shape of things. The hidden order.

Looking back, Benoît recalled: "My first work after prices happened to be of a very different kind. It was about errors and telephone channels. This was something that IBM was worrying about very much. Until this time, computers were big boxes in a room with some data. Then the idea of terminals began and thereafter the idea of transmitting data over telephone lines — ordinary telephone lines. It was done in a fashion which people felt was very conservative, but in fact it had very complicated and strange results. A friend of mine, listening to me speaking about financial prices said, 'Well, it seems to me that what you've been doing may, perhaps, help you understand the mess which I'm in.' I suggested to him to do a few tests."

The results of the tests were very positive. Benoît suggested to his friend that he could inform his bosses that their technology was just stupid and outdated. It was very costly and ineffective. They were trying to do the impossible at great cost. At his suggestion, they abandoned that old technology. It was also abandoned by Bell Laboratories and AT&T after Benoît had given a presentation on his findings. A totally different technology was then introduced which was appropriate.

Martin Gutzwiller: "For some 20 golden years, the big corporations decided that they should treat science as important, just as they used to do in the 18th century. It was when scientists could add to their glory, and gain a reputation as reasonable and scientifically orientated people. The important people in the large corporations had the same kind of benefit and they were hoping, of course, that some people would actually use some of their minds to solve the problems that the businesses were facing."

04

THE BIRTH OF FRACTAL GEOMETRY

Plato sought to explain nature with five regular solid forms. Newton and Kepler bent Plato's circle into an ellipse. Modern science analysed Plato's shapes into particles and waves, and generalised the curves of Newton and Kepler to relative probabilities — still without a single 'rough edge'. Now, more than two thousand years after Plato, nearly three hundred years after Newton, Benoît Mandelbrot has established a discovery that ranks with the laws of regular motion.

Professor Eugene Stanley

© Gabriel Lesmoir-Gordon

The problem of the transmission of data over telephone lines at IBM was a most interesting case for Benoît. In order to understand that 'messiness' with the phone lines, Benoît used the intellectual tools which were exactly the same as those he had used to look at messiness in financial prices. They were also the same tools that he had learned as a student, and which had been introduced into mathematics a hundred years before as so-called 'pathologies'.

Benoît had looked back to the masters of mathematics at around 1900. At that time, objects like the Cantor Set, the Sierpinski Triangle, the Koch and Peano Curves were known by a few, but pure mathematicians were mostly the ones who were convinced that those objects or shapes proved that pure mathematical thought was quite separate from reality because they thought that those ideas had no practical implementation in nature.

Benoît found that to be quite the contrary. "Those tools were precisely the tools, which I needed to understand the *holy mess* that I was encountering in many areas." He grasped that the noise in the IBM system was in fact deeply embedded in nature and impossible to drive out. He instantly doomed any attempts to predict, suppress, or eliminate it. This noise issue was an early and very clear example of the strange logic of fractals — 'that unruly collection of irregular geometric phenomena' that only Benoît comprehended at that time.

The Polish mathematician Vaclav Sierpinski introduced his fractal in 1916, but the underlying principles were known to artists for centuries. The Sierpinski Triangle or Gasket is obtained by starting with a filled equilateral triangle, which is then divided into four smaller equilateral triangles, of which the middle one is removed, leaving a triangular hole. The three remaining filled equilateral triangles are then divided in exactly the same fashion so that three smaller triangular holes appear. Conceptually, we can repeat this process indefinitely at smaller and smaller scales reaching, in the limit, Sierpinski's Gasket, a shape composed of three copies of itself, each half as big as the whole.

The pulpit of the Cathedral of Ravello contains panels which feature the Sierpinski Triangle. The pulpit was created in 1272 by sculptor Nicola di Bartolomeo da Foggia.

This rule applies equally to squares, pentagons or any other polygons. Similar techniques work with circles too.

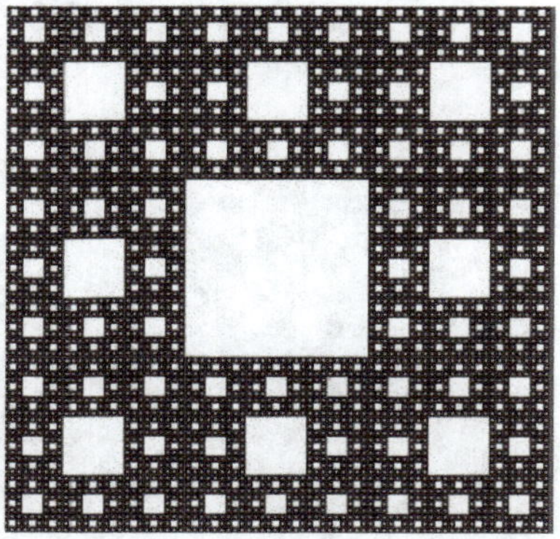

The Sierpinski Triangle

Building up the Sierpinski Triangle

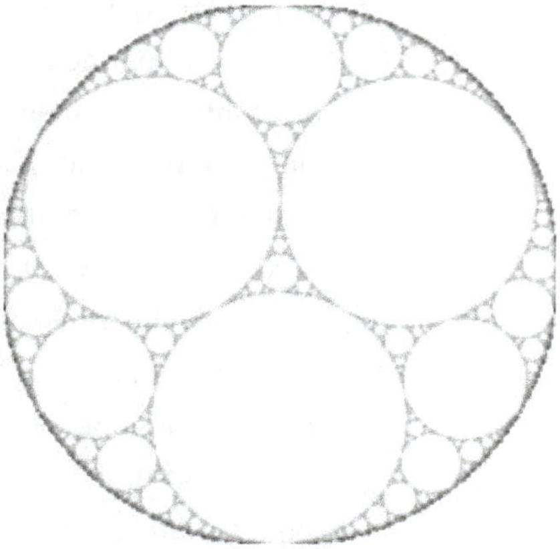

From the pulpit of Ravello Cathedral

These patterns made by successively carving three circles out of ever-decreasing circles show remarkable similarities to Celtic art. The Book of Kells is a graphic example. The book was created in a monastery on the Isle of Iona in Scotland in the early 8th century to honour Saint Columba.

Book of Kells

The mathematician Heinz-Otto Peitgen concedes that: "Now we know that there were a lot of what we now call fractals at the turn of the last century. But all these things like the Koch Curve, the Cantor Set, and the Sierpinsky Triangle, were mathematical topics and mathematical shapes which were not seen in relation to nature at all. They were seen as examples of a particular and very peculiar mathematical property for which they were made at the time. They were certainly not made to explore nature as such. They were made to explore mathematical questions very far removed from nature.

© Gabriel Lesmoir-Gordon

I think it took the big leap which Benoît made. It takes quite a leap to see that these forms have something in common with nature. Of course, we now know through Benoît's work that they have more than just something in common with nature. They really are at the very heart of understanding the patterns, the structure, the irregular shapes of nature in a very deep way."

Georg Cantor's quest for the meaning of continuity led him in 1883 to the set that is now named after him. It was one of the first fractals to be studied mathematically.

The Cantor set begins by taking a line and removing the middle third leaving two equal lines. Likewise remove the middle thirds from each of these two lines. Repeat this process an infinite number of times and you are left with the Cantor set.

An explicit formula for the Cantor set is:

$$C = [0,1] - \bigcup_{m=1}^{\infty} \bigcup_{k=0}^{3^{m-1}-1} \left(\frac{3k+1}{3^m}, \frac{3k+2}{3^m} \right)$$

The Cantor set has no length or interior. In technical parlance, it has zero measure. A randomly thrown dart is infinitely unlikely to hit it.[1] The existence of this set forced Cantor to refine his notion of continuity. Mathematicians were searching for a definition of dimension.

[1] It is nowhere dense. Every part of it consists almost completely of holes. And yet, despite being nothing but totally disconnected points, it is uncountable.

A bombshell burst around 1890 when Giuseppe Peano discovered what was called a 'space-filling curve'. Peano had constructed an idealised curve which twisted in such a complex way that it visited every point in the entire plane.

If you tried to plot it, it would actually fill the whole plane of the paper it was drawn on. There was no point on the plane that Peano's curving line would not include.

The Peano Curve

This led to a precise definition of the dimension of a shape or space called the topological dimension. However, another line of reasoning by Felix Hausdorff led to a different take on dimension.

For normal Euclidean objects, the two definitions were in exact agreement. Both gave the same whole number. For fractal shapes however, Hausdorff's approach gave a fractal dimension, which allows for the intriguing possibility of one-and-a-half-dimensional objects.

These shapes are incredibly complicated to describe in Euclidean terms, yet share an affinity with many shapes of modern mathematics displaying an endless series of motifs within motifs repeated at all scales.

One such shape is the snowflake curve, devised by Helge von Koch in 1904. He defined the curve as the limit of an infinite sequence of increasingly wrinkly curves. The finished curve is infinitely long despite being contained in a finite area. It has no tangent or smoothness anywhere. Slicing the curve at certain angles reveals an infinity of

The Koch Curve

Cantor sets lurking within. What von Koch did not realise, however, was that such curves with infinite length would make ideal models for shapes in the real world like coastlines.

Heinz-Otto Peitgen: “What I find amazing about this background, which is relevant to fractal geometry, is that it’s hidden in many places; it’s not just in one particular mathematical field. It’s in probability theory. It’s in classical geometry. It’s in what we call topology. It’s in many places and most of these places, which were developed late in the 19th century and early in the 20th century, were more or less forgotten. This doesn’t mean they were unimportant at the time. They were in fact extremely important when they were developed, but they were forgotten. What is so interesting about Benoît is that he somehow had this incredibly broad knowledge of mathematics which goes back into history. He seems to have known it all when he became aware of how these different strands, and how these different forces, were playing together in a new way. That’s what makes him unique. This was the beginning of fractals. The principal criterion used to model those data were in point of fact an aspect of folklore.”

Benoît: “The engineers at IBM were very upset by the fact that sometimes their data went down due to individual errors and they knew that the errors were clustered in bunches. There would be an error then there was no error for a long time, then another bunch with no error, then another bunch with an error. If they looked at microseconds in the error, it looked exactly the same as when they looked at seconds in the error. This terribly complicated phenomenon had the strange property of having the same appearance whether looked at very, very close or farther away and farther away. When they zoomed in or zoomed out the phenomenon did not seem to change. This was not viewed as being a major discovery — it was a bit of folklore — but that bit of folklore was the key. On one occassion, a professor came to give a talk about turbulence at Harvard where I was a visiting professor. The story of that visitor was a tale of woe. He said he had this data: "It's a mess. It's so incomprehensible. I can't understand anything. I tried this and it doesn't work. I tried that and it doesn't work. I tried something else and it still doesn't work."

He explained to Benoît how terribly difficult the problem was for him. He told him later that he saw him sitting in the first row 'just beaming'. Benoît was 'smiling from ear to ear' because he knew exactly what to do with this problem.

"It required the same intellectual technique, the same way of reaching into those mathematical esoterica which nobody thought were real. I could transform the issue on the spot, not immediately solving his problem because the problem was very difficult, but making sense of it. All those contradictory results which he had were in fact not contradictory but part of the basis of the model which I had produced for turbulence."

Turbulence was another thing Benoît had learnt about at Caltech. He did not know then that he would ever come to apply his turbulence theory in this way. But he was well informed about the way that turbulence and fluids behaved which is why he could so easily and rapidly jump straight into this problem, following on from his work on financial data and telephone errors.

"How could I tell people that I was using these horribly abstract notions which everybody told them were of no use at all in science, were the required tools for this data? Now people were not very happy with it for this very simple reason: their lack of experience. I was not accused of being a charlatan or a fraud because everything *seemed* to be right. The problem for me then was: how to explain it, how to make it natural as opposed to being unnatural, crazy and wild."

During the 1950s, Benoît's quest took him on to the stars and the study of galaxy clusters where he applied his ideas on scaling to the overall structure of the universe. He was able to show how matter in the universe is distributed in such a way that the density decreases as the volume increases. It is fractal.

The work of Luciano Pietronero also showed that the distribution of mass in the universe is fractal. Up to about ten years ago, astronomers always maintained that the fractal nature of the universe was cut off at a particular length-scale. Pietronero looked very carefully at atlases of the universe. His objective empirical measurements, using fractal geometry, led him to conclude that the distribution of mass throughout the entire universe is probably fractal.[2] And then the question was: where does this fractal structure come from?

[2] Some astronomers have reached a different conclusion about the distribution of galaxies.

Most of the physics, which we try to learn in school, concerns itself with phenomena which have a well-defined scale. For example, a pendulum has a period which is very well defined. The planets go around the Sun and each planet has two scales: the size of the planet and the time it takes to go around the Sun.

They move by fixed rules which are very important in physics. Each of these phenomena has a well-defined scale. Over the first half of the 20th century and beyond, physicists became obsessed with phenomena which did not have a very defined scale. For example, the main property of turbulence is a combination of eddies of all sizes — big eddies, small eddies and very small eddies. This is something which has been known to engineers for a long time. Leonardo da Vinci was the greatest of all the medieval engineers. He made drawings of fountains which show big eddies, middle-sized eddies and small eddies writhing over each other. It is impossible to explain and describe turbulence by looking only at the big eddies and forgetting the rest. This is the reason behind the immense difficulties encountered in weather prediction. We cannot possibly deal with the major elements of the weather — the big eddies — while overlooking the finer details. The details affect the big things, and conversely. The study of turbulence has encountered enormous difficulties dealing with this issue. It is a field which does indeed continue to be very baffling.

A similar phenomenon occurs, for example, in magnets and in other areas of statistical physics which do not have a single, well-defined scale but have a variety of different scales. Now that is the part of physics where Benoît's work has had the most influence.

He introduced geometric aspect into physics, concluding that physics is ruled by geometry — that physics and geometry are intimately linked. He maintained that one cannot do the physics without understanding the geometry.

Martin Gutzwiller wrote one paper with Benoît where their views converged nicely while dealing with the same problem independently. When they talked to each other, it occurred to Gutzwiller that he was doing something along the same lines as Benoît, and that he needed an answer from him. That answer had to do with the fractal nature of an object which he was interested in but did not know the technical requirements for stating it properly. It was a question that Gutzwiller had been busy with for 35 years or so and which people thought was well-understood. It connected in with what is now called Chaos Theory. Gutzwiller was interested in how chaos manifests itself on the atomic scale. Benoît explained to him how to approach the subject. Gutzwiller did the calculations and in the end Benoît was happy with what he had achieved but he did not quite trust it. He asked a graduate student to do it over again. Eventually it was straightened out.

Acting like an 18th century naturalist, Benoît scoured through forgotten and obscure journals in his quest for insight. Fortunately, he uncovered the work of an eccentric and unremembered mathematician called Lewis Fry Richardson. Benoît had struck a rich seam and he knew it.

Then a very fortunate thing came up for Benoît again from a waste basket, or more precisely from a pile of material which was being disposed of. The library at IBM was discarding a vast quantity of books which nobody had ever looked at because there was no room for them. They could not keep them and they decided that to keep a book for ever and to have a bigger and bigger library was not possible. Benoît fortunately was on very friendly terms with the librarian and she regularly told him when a truck was going to come and pick up books for pulping. Benoît would look through them. Most of them were the proceedings of meetings that nobody was at all interested in and were only suitable for pulping. But he came across one periodical however which he had opened more-or-less at random and there he saw the name Lewis Fry Richardson. This name was already very well-known to him through Richardson's work on turbulence in the 1920s. Richardson was one of Benoît's great heroes. He was a very strange character. He was a very great man in many ways but a true English eccentric.

Richardson had measured the length of coastlines at different scales. There was one picture in that periodical that really caught Benoît's attention.

Lewis Fry Richardson attempts the high jump

He took a Xerox copy of the picture and set the periodical aside to pick it up later. The next day it was gone and he only had the one picture, but that picture was one of the keys to fractal geometry because it referred to something we all know something about — coastlines.

Richardson delighted in asking questions no one else even considered worth asking. One of his papers from 1961 entitled *Does the wind possess a velocity?* anticipated later work by Edward Lorenz and the other founders of Chaos Theory.

One of Lorenz's great insights was a model of turbulence as a collection of ever-smaller eddies. He conveyed the idea poetically in the style of Jonathan Swift:

"Big whorls have little whorls, which feed on their velocity; and little whorls have lesser whorls, and so on to viscosity."

What really struck Benoît forcefully were Richardson's 1961 ideas on the lengths of coastlines. What appeared to be a simple question of geography transpired, on further consideration, to expose some of the essential features of fractal geometry.

When Benoît attempted to explain to economists that prices were discontinuous, it made them angry and defensive. Their response was that they could appreciate that prices change very fast, but not that they were discontinuous. Benoît tried to explain to people some exotic facts about turbulence. "They always knew too much and they had their own answers to the question. Half answers in fact. On coastlines, however, people were neutral because they could understand that if you take a measurement of the coast of Britain, say on a sphere, it is very, very rough. And Wales is just a bump and Cornwall a bump. Scotland is very smooth. If you take Europe you see Wales is not so smooth, that Cornwall has jiggles and Scotland is anything but smooth. The closer you come, the longer the coastline becomes. If you just look at a coastline with your eyes you can see it is very seldom straight. A beach can be straight but a real coastline can never be straight. So you must expect to have a length that increases as you come closer. And that's where the dimension — the fractal dimension — came in. As a result I was telling my friends that this is the new way of looking, for example, at errors in telephone systems in terms of their dimensions being equal to one-third. But engineers said it was ridiculous and that I was just making it all up to impress them. Mathematicians were saying: 'You're ridiculous because dimension has no application of any practical nature!'"

As far as coastlines were concerned Benoît found that his theory worked beautifully. He saw that people frequently fled in horror if the idea was misrepresented, but if it was represented in the proper way they immediately accepted it. This paper in fact became very, very famous. The title was *How Long is the Coast of Britain?*

Richardson coined the phrase and now it is something we use all the time.

Benoît chose Britain as his example because he appreciated what this could do for Britain in English and Brittany in French! It was an easy, nice and dual-purpose phrase. His article on the coastline of Britain started him off in a serious way on his fractal enterprise which was to go on for many years thereafter enhancing our understanding of financial prices, the fluctuations of dissipation in turbulence and in the distribution of galaxies. In these and many other areas, Benoît's procedures started in much the same way. The beginning was always the same, however, because he knew exactly how to deal with 'various kinds of messes'.

Each new application required that he studied something new because one thing Benoît never did was to 'come, burn and flee'. He never went into any field without being well informed about it. When he had some new idea, he went into the field boldly. He always wanted to know the 'folklore of the field'. Very often he found that the folklore told him much more than the analysis. In the case of the phone line errors, for example, where the errors looked the same at all scales. That was folklore, but it reminded Benoît of the idea that Richardson had for turbulence — that turbulence was the same on all scales but which Richardson had not actually pushed very far. Benoît pushed the idea much further. "In each case the same general idea applied: that bodies are very complicated, very irregular, very broken up and that the irregularities are the same at all scales."

In 1973, Benoît started to develop an algorithm for creating real looking coastlines with Sigmund Handelman at IBM. Handelman used a very basic, make-shift computer with a minute memory to generate the first pictures which imitated natural landforms. The tools like self-similarity — the fact that certain objects are the same at all scales — became the most common ideas behind fractal geometry which at this stage though was a field that was more or less outlined in Benoît's mind but still did not exist outside it.

Then computer graphics started to become a reality. This was an extraordinary and highly important event for Benoît. The only place then where they had computer graphics facilities was at Los Alamos and some other military research establishments. Benoît did not have access to these facilities. Although IBM was not in the business of developing computer graphics systems, Benoît was able to develop a make-shift CG set-up.

To begin with, using typewriters, they were able to get a computer to fake a coastline which fitted his theory and satisfied Richardson's empirical results. Benoît recalled: "We had to sit in front of a typewriter and eventually that typewriter would type the outline of the coastline. Now that was horrible. It is unimaginable now. Someone had written a program in which the memory was 64 bytes. Not 64 kilobytes, not 64 megabytes, just 64, period. It was a tiny memory. Each time that memory was exceeded we had to push return so it would add another 64 more bytes. I had this totally minute, ridiculous programme to use for this ambitious project to make coastlines. I spent a night with my assistant sitting there pushing and pushing return again and again. Each time a copy stopped he pushed return and it started again and eventually we got the coastline. I filled them in with a black marker to produce maps of different imaginary continents."

Benoît continued to work on a variety of apparently unrelated problems. He returned to economics in an attempt to make sense out of fluctuations in the prices of commodities like cotton which he found and which exhibited what he later called the Joseph Effect[3]. His model seemed to fit the bill and was very effective in imitating the apparently random fluctuations of nature. He then moved back into the study of turbulence.

The number of topics Benoît was prepared to handle was increasing rapidly because he was becoming more and more confident by the day. Looking back, he said: "I was successful about half of the time. Half the time it was ridiculous and when it did not work, I just dropped it.

I never had the impression that I was finding the cure for everything in the universe. But the key which I was developing opened very many different doors, to my and every one's surprise."

His work became focused in the winter of 1972–1973. He had been away from France for a while and returned to Paris for a sabbatical. He gave many lectures, including at the College de France. "This was one of the harshest examinations I had ever taken in my life because I was returning to Paris after 15 years of absence. People had heard of my work in finance, my work on mountains and so on, which was totally incoherent to them. Moving from one field to the other seemed to be a mark of defeat because nobody abandons something that works well to start something that has not even started. And there was a very great curiosity about my lecture at the College de France. Since that lecture was held on a Saturday it attracted a very large audience.

I prepared my lecture very, very carefully. That task was highly constructive. I never before had the experience of a lecture forcing me to such feats of reorganisation. And fractals were in a certain sense born for this lecture. That is, the listing of the techniques, the search, the scope of the ideas and so on, but the name did not come yet.

The topic I was studying still had no name, because I was the one to conceive that such a subject — the study of irregularity, roughness, 'broken up-ness' — was a real possibility. Not on the level of general impressions but on the level of very sharp, close mathematics and in being able to make imitations. This was my argument:

[3] The idea that movements in a time series tend to be part of larger trends and cycles more often than they are completely random. The Joseph Effect is quantified by the Hurst component, where movements fall between a Hurst range of 0 to 1. The term was coined by Benoit.

© Gabriel Lesmoir-Gordon

if you understand something, you may be able to imitate it. If you cannot imitate it, there's something lacking in your understanding."

Benoît set about organising his ideas and putting them down on paper. He was preparing to publish his first book in France in 1975. The book was not going to be very ambitious. It was just 200 smallish pages. He did not see it as a serious textbook. He simply set out to explain what he had been doing, what he had discovered. When the book was close to completion, he was challenged to give a name to his creation and he coined this word *fractal*. One day he was thumbing through his son's Latin dictionary, idly looking for a word he might adopt. He came across the word *fractus*. "Why did I choose it? Because I had a classical education. I had Latin and I wanted something which presented a notion of broken up, fragmented, irregular, interrupted. The Latin word *fractus* is precisely the right one for it. It's like when you kick a stone hard it falls into pieces and the pieces are very irregular. The word *fractus* has given fraction, which actually does not have the notion of irregularity at all, but the notion of parts. And fracture, which does have the notion of irregularity. So, I adopted the word *fractus, fracta, fractum* and coined the word *fractal*."

Martin Gutzwiller: "Benoît introduced the word fractal to the world and he even had a grammatical explanation for it: how the plural would be, and so on. The objects did exist before in some kind of paradoxical, intellectual constructions which were highly artificial and therefore were not taken seriously. Those few mathematicians, who knew of them, dabbled around and then forgot about them. But Benoît picked them up and saw immediately that nature has

chosen this particular feature as a worthwhile technical help in accomplishing whatever nature tries to accomplish. Benoît saw that very, very clearly."

The coining of a name had an extraordinary and immediate effect. It was no longer the crazy collection of Mandelbrot's *miscellania,* but was the study of a group of objects. A name was a form of existence.

The physicist Bernard Sapoval says: "The creation of a word is extremely important. There are not so many words, so if you are able to add one word, it's a big thing! And it's a very significant word."

Though the idea of fractals in the form of iteration and self-similarity is an ancient one, it took this wanderer-by-choice to give the idea a name. Once named, the field took off. As Benoît put it, 'to have a name is to be'.

The Romanesco Cauliflower is an excellent example of a fractal structure that goes in and in and in and where the same pattern is repeated over many scales. If a section is broken off it looks like the whole thing, and if a section

A Romanesco Cauliflower

Ferns in nature

A computer-generated fern

of that is removed again it looks like the whole thing. Another example: little pieces of a fern look like the whole thing.

Benoît maintains that his book of 1975 and the creation of the word *fractal* finally allowed his field a reality. He emphasised that though the field had existed as a broad collection of activities, none of them was viewed as being a major, standard, successful part of science. There was very little new mathematics in his book of 1975 and there was very little physics in it either. When he wrote that book he was in the process of creating something which he felt he would never fully achieve. But he had discovered a construction for fields as varied as turbulence and finance. It needed development. Benoît approached this work very much in the spirit of his uncle Szolem.

Carl Everstz: "At that time most of the people, who read *The Fractal Geometry of Nature* acknowledged that after reading the book they simply did not see clouds anymore in the way that that they saw them before. They did not see trees as they saw them before. Many things changed."

Michael Barnsley: "Once we develop a fractal geometer's eye, we see them everywhere. Every single thing we see in this world is described by reference either to itself or to something else.

Michael Barnsley

It's as though we are looking at a vast dictionary and the dictionary words are bits of pictures and the definitions of the words are made with other bits of pictures."

Ian Stewart

Ian Stewart: "Living creatures seem to be complicated structures produced from simple rules, simple laws of physics and chemistry. A lot of the structure that we see in living creatures — this organic, but patterned structure — ferns, leaves on trees, particularly things like that — have the same feature that the Mandelbrot set has. We look at little pieces of them and they have lots and lots of detail. The little pieces look very similar to the whole thing."

Arthur C. Clarke: "As we look closely at Saturn, we see more and more detail which no one had ever dreamed of before the space age opened and we were able to get close-ups of Saturn and its rings. It's not surprising that when we have so many examples of fractal and related phenomena here on this planet that there are even more in the heavens."

Michael Barnsley: "To me, just looking up at the Milky Way is staring at a fractal. It's got an extraordinary dotty character and yet if we take a telescope and look at it ever closer we find that there are hundreds, thousands more little dots where we thought there were almost none. So, we have another example of a structure that goes in and in and in with more and more detail."

Arthur C. Clarke: "I once had the great privilege of having a discussion

with the cosmologist Professor Stephen Hawking. I said to him," The Mandelbrot set is infinite in detail. You can explore it forever and ever — zoom into it. The real Universe does seem to have limits though. As you go down into the micro-world you get molecules, atoms, neutrons and sub-atomic particles — quarks. But does the real universe go on forever, is there a limit — a basement — unlike the Mandelbrot set?

Professor Hawking's response was that in the case of the universe there seems to be a limiting scale and it is called the Planck Length which is about a million, billion, billions, billion times smaller than a centimetre. This means that there is a limit to how complex the universe can be. It also means that the universe could be described by a theory that is fairly simple, at least on scales of the Planck Length. So Hawking takes the view that this length is the fundamental unit of size — the sort of grittiness — of the universe. There is nothing smaller than that. So perhaps the real universe does end there in smallness. But we're not sure. It may indeed go on forever like the Mandelbrot set. We simply don't know yet."

05

THE MANDELBROT SET

On 1 March 1980 at IBM Yorktown Heights in upstate New York, Benoît discovered what is now known as the Mandelbrot set. The irony is that Mandelbrot's uncle Szolem had strongly suggested that Benoît look at the papers that Gaston Julia and Pierre Fatou had published in 1917 with a view to making them the subject of his PhD. Re-examining the maths behind the Julia sets again in 1980 — and this time on a computer — led Benoît directly to the discovery of the Mandelbrot set.

Benoît: "We come now to a year which was an almost miraculous year in my life. It was in 1979/80 that my second major contribution to mathematics took place and in which my major contribution to *almost* mainstream physics took place. It was the year that I discovered the Mandelbrot set — probably my work that everybody knows best. But it came in a very, very slow, long and drawn-out fashion. When I was struggling with the question of *who am I?* my uncle Szolem thought that I should know about the theory of the Iteration of Functions which means that you take a number then a function of that number, then a function of that function, and a function of that function. Iteration is doubling. You take two and four and eight. You double each time. But it's not just doubling. In fact it's a more complicated operation. That theory had been a triumph of major proportions in 1917 and 1918 when two French mathematicians, one of whom had been one of my teachers named Gaston Julia and another named Pierre Fatou."

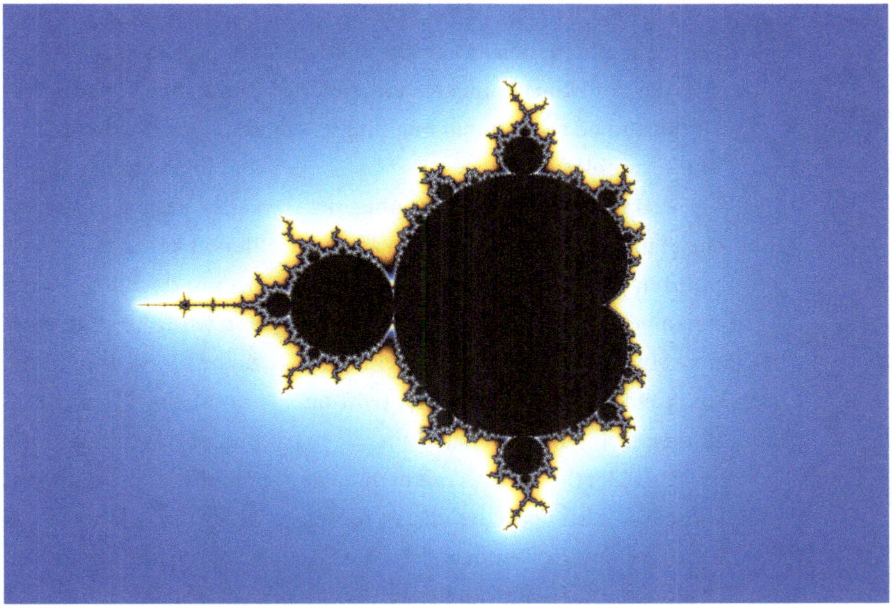

The Mandelbrot set

Complex Numbers have their own arithmetic, algebra and analysis.

They rely for their existence on an act of pure mathematical imagination: that is, to agree that minus 1 is allowed to have a square root.

$$\sqrt{-1}$$

Complex Numbers have two components or co-ordinates 'real' and 'imaginary'. The imaginary component contains the factor **'i'**, which stands for the $\sqrt{-1}$.

It is this **'i'** which has opened up the world of the Mandelbrot set.

Since the square root of a negative number cannot be placed anywhere on the number line, mathematicians up to the 19th century could not ascribe any sense of reality to those quantities.

The great Gottfried Leibnitz, inventor of the Differential Calculus, attributed a mystical quality to $\sqrt{-1}$, seeing it as a manifestation of the Divine Spirit. He called it *'that amphibian between being and not being'*.

A century later, Leonhard Euler wrote in his work ALGEBRA in words that still echo the same sense of wonder:

'All expressions such as $\sqrt{-1}$ are impossible or imaginary numbers, since they represent roots of negative quantities. Of such numbers we may truly assert that they are neither nothing, nor greater than nothing, nor less than nothing.'

During the First World War the French mathematicians Gaston Julia, a student of Henri Poincaré, and Pierre Fatou studied the rational mappings[4] of the complex plane.

[4] Ratios of polynomials

A transformation or mapping of the plane is a rule which, given any point in the plane, provides another. It can be thought of as acting simultaneously on the entire plane: picking it up, stretching it, spinning it, or twisting it, then laying it down flat again.

In 1685, the English mathematician John Wallis hit upon the idea of representing complex numbers graphically in a diagram. The imaginary numbers are strung out at right angles to the line of real numbers, also known as the real line. If we portray real numbers as lying on an east/west axis, then imaginary numbers lie on the perpendicular axis through the point zero and run from north to south. This creates a cross coordinate system where all the real numbers are placed on one axis and all the imaginary numbers on the other with a general point in the plane consisting of a real part and an imaginary part. Arithmetic operations on complex numbers correspond to geometrical transformations of the complex plane. Addition is equivalent to translation; multiplication to scaled rotation.

Benoît had a classical training in mathematics. He also had a phenomenal memory. Everything that went into his head stayed there and he always maintained access to it. From his classical mathematical background, he had all the information which he required about the work of Julia, Fatou, Sierpinski, Peano, Cantor and Koch. These men constructed curves and shapes that other mathematicians, and in particular Henri Poincaré, had called pathological monsters. But Benoît kept them at the forefront of his mind while he was working at IBM and while he considered 'roughness' as the foundation of fractals.

Michael Frame: "When the computer came along, he realised that he could actually draw pictures of these things. When he saw them, and he was a very visual thinker, the floodgates opened. Millions of things came out and that was just the beginning."

Julia and Fatou had also looked at the process of iteration. Without modern computer graphics, it was almost impossible to visualize their subtle ideas. Self-similarity, for example, was well known to Julia and Fatou. The mappings Julia studied appeared to be discrete, yet various analogies with real continuous dynamical systems arose naturally.

The two mathematicians found attractors, points in space, which pulled all the surrounding points towards them. And they found repellors too. These created basins of attraction much like falling rain has to find its way into one river or another. The boundary between the catchment areas is made up of repelling points, which push neighbouring points away. These boundaries turned out to be very complicated. They are now known as Julia sets although Julia and Fatou never actually

Michael Frame

saw a Julia set. The first attempted visualisation appeared in a paper published in 1925. This was the closest they ever came to seeing the full flowering of their work. It was not until the advent of modern computers that Julia sets could be seen for the first time in their full glory and with all their detail exposed.

Hidden behind the apparent diversity of the Julia sets a powerful unifying principle was waiting to be discovered: a key to untold mysteries, a sleeping dragon. And only one person was radical enough to realize its importance.

Ivar Giaever

Ivar Giaever: "There existed something before, which people called Julia sets, but since they really could not plot them out except by an enormous amount of work, they were really never looked at. So Benoît was lucky that he came in when people were starting to show pictures on computers. The timing was perfect. No question about it."

Benoît: "Julia and Fatou wrote a paper. It was a huge idea even though it was a short paper. People didn't know what to do with this work. My uncle always said that it was a beautiful theory, but nobody had done anything about it for 30 years and that was back in 1947. 'Why don't you look at it?' he asked me. 'You could do better. You could go back into the mainstream with this and on your own terms.' But I couldn't see anything to do with it. It was one those perfect egg-shapes — everything that they could think of had been thought of. But they did not have the computer. So again, 30 years later, I came back to the same problem. I was a geometer in every way. I was thinking of those shapes, but all these shapes were purely in my head. Following my work on mountains, I could see for the first time exactly what was in my head, show it to others and use it to better and better advantage. I went in that direction with a vengeance. I returned in particular to that very old problem in mathematics — very pure, very abstract mathematics — using pictures to see what those theorems really meant. And that wasn't so simple. The computer can do pictures of complication way beyond what the hand can do, what the mind can only imagine — and I was very good at imagining them. I could actually see what they were, look at them and modify them. I found myself then feeling strongly that I was like an 18th century natural philosopher. For me, the first step with any difficult mathematical problem was to program it and see what it looked like. We started programming Julia sets of all kinds. It was extraordinary great fun! At one point we became particularly interested in the Julia set of the simplest possible transformation, that is *Z* goes to *Z* squared plus *C*. So *Z* times *Z* plus *C*."

This formula gives a rule for getting one complex number from another, or in other words, mapping the complex plane onto itself. The effect of this mapping is to cut the plane, wrap it around itself while stretching it away from the unit circle. Benoît called the Julia sets generated by this formula *self-squared dragons.*

The Julia sets for this mapping depend only on the value of the parameter *c*. When *c* is small, they are simple loops like wrinkled circles.

For large values of c, the fractals consist of innumerably many discrete points, spread out and dust-like.

Generally, the Julia sets can be divided into two main types, two varieties. They can be either wholly disconnected and dust-like or wholly connected. In the former case, the sets are topologically the same as the classic Cantor set.

With the connected Julia sets, on the other hand, each consist of a succession of lines. Sometimes a single closed curve; sometimes loops within loops within loops; and, from time to time, a dendrite.

On the boundary between these two regions are dendrite Julia sets, made up completely of continuously sub-branching lines which are only just connected since the removal of any point from them would split them in two.

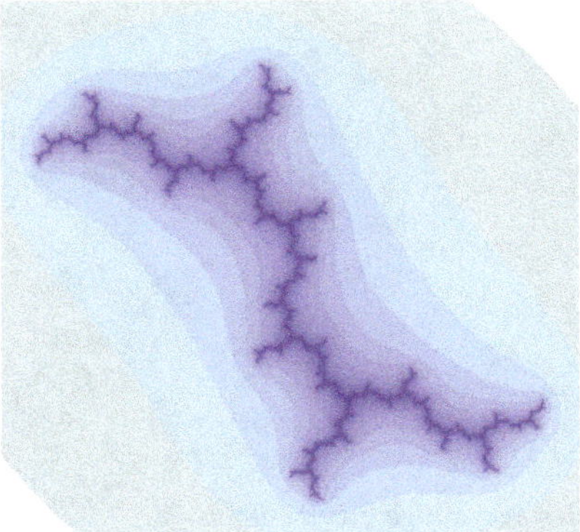

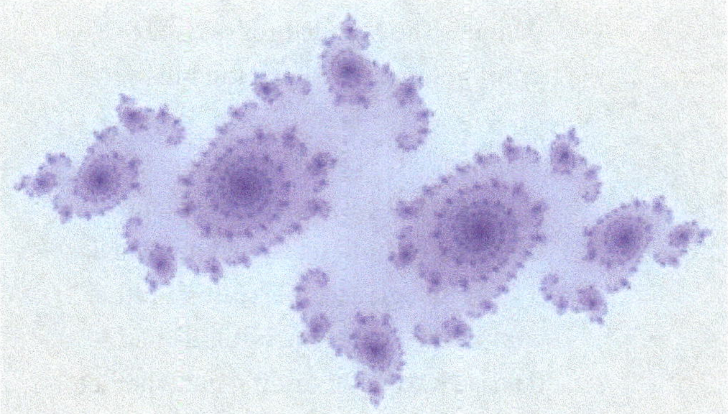

© Michael Frame

Benoît started looking for the pattern behind this dichotomy, the index to the book of all Julia sets. The questions he asked were: 'What is the grand plan behind this behaviour?' and 'What separates the joined-up Julia sets from the dust-like ones?' He and his team spent a long time churning out Julia sets.

In 1980, Mandelbrot hit upon the fruitful idea of making a map of this behaviour. He coloured a point on the map black if a Julia set from that location is connected and leaving it white if the Julia set is disconnected. This is how the Mandelbrot set is defined: The Mandelbrot set is the set of points c for which the Julia set of $Z > z^2 + c$ is connected. If c is in the Mandelbrot set, the Julia set is connected. If c is outside the Mandelbrot set, the Julia set is not connected.

At first glance, this seems a rather complicated problem, but Julia had devised a trick for finding out whether or not a Julia set is connected without actually constructing the set itself. If the iterates starting from $z = 0$ are examined and if they go to infinity, then the set is disconnected, otherwise it is connected. With this insight, a simple computer program can find out to which class a Julia set belongs.

When the first picture of the set rolled off the printer, Benoît and his colleagues' first reaction was that there must be some mistake, a fault in the program perhaps. The picture was extremely strange and very unexpected.

Over a period of weeks, working late into the night in a basement laboratory Benoît and his assistant explored the astonishing new world they had discovered. By feeding new coordinates into their program, they made successively deeper zooms into the boundary of the set. One of the most striking discoveries was that buried deep within the seething froth of the object's boundary were tiny replicas almost identical to the original set. Zooming further into these baby M-sets, they encountered further variations of these very same patterns

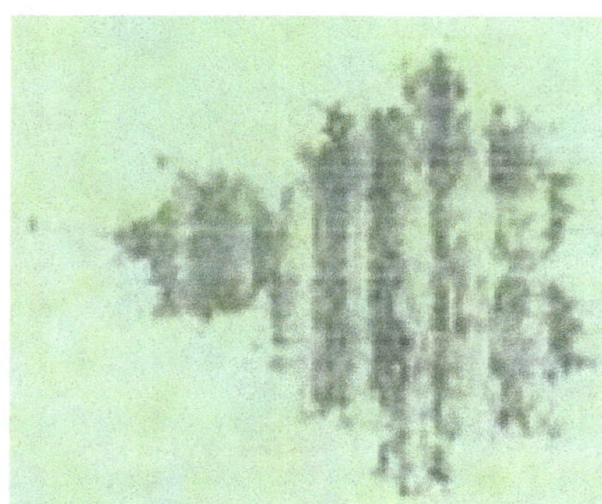

The first very rough picture of the Mandelbrot set

with added frills and embellishments. These baby M-sets were like a key, revealing the existence of a whole new world of embedded Julia sets. These consist of an infinite number of scaled-down copies of one Julia set, strung out on spiral strands in the general shape of another.

The most interesting Julia sets are the ones which are only just connected. These lie on the border between the connected and the dust-like sets — that is on the boundary of the Mandelbrot set. Likewise, the Mandelbrot set is most interesting at the boundary. The Chinese mathematician Tan Lei proved that the M-set is asymptotically similar to Julia sets at almost every point on its boundary. What this means is that all the patterns found in these Julia sets are also found on the boundary of the Mandelbrot set. The more you zoom into the Mandelbrot set, the closer it resembles some particular Julia set. For this reason, the M-set has been called a pictorial index of Julia sets.

"I made many pictures of it. The first one was very rough. But the very rough pictures were not the answer. Each rough picture asked a question. So I made another picture, and another picture. And after a few weeks, we had this very strong, overwhelming impression that this was a kind of big bear we had encountered!"

Benoît and his team pushed the computers of the day to their extreme limits to refine their images. Then their luck broke. Their pictures instead of becoming increasingly sharp became increasingly messy. They took a closer look. Many specks of dirt vanished as they zoomed in, but some did not. They resolved in fact into structures similar to the main set. Peter Moldave, Benoît's moonlighting research assistant, could not contain his excitement.

Ian Stewart: "There is an interesting parallel here with the equation that almost everybody is familiar with, the only equation that everybody is familiar with: $E = mc^2$. Albert Einstein's equation, which states that matter and

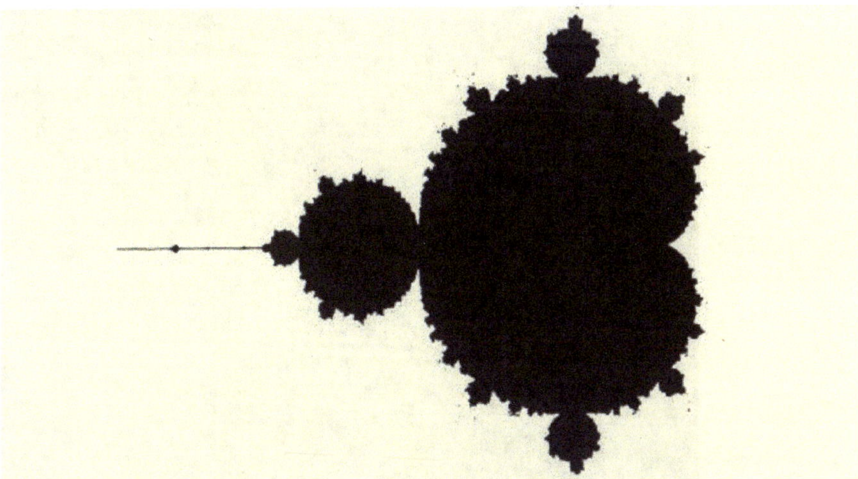

An early but more refined image of the Mandelbrot set

energy are equivalent to each other. That was a very simple equation but with very far-reaching consequences. And the equation for the Mandelbrot set is equally simple: $z \longleftrightarrow z^2 + c'$"

Arthur C. Clarke: "The letters in the Mandelbrot equation stand for numbers, unlike those in Einstein's equation where they stand for the physical quantities: mass, velocity and energy. The Mandelbrot set equation numbers are coordinates, positions on the plane, defining the location of a spot. Another difference from Einstein's equation — and a very important one — is the one-way arrow in the middle:

$$z \longleftrightarrow z^2 + c_i$$

When the Mandelbrot equation is given a number representing a point and that number is iterated through the equation, one of two things happen. Either the number gets bigger and bigger and runs out to infinity or it stays inside a circle of radius 2. Depending on which happens, the computer then knows where to draw a boundary line. So, what is obtained from this basic iteration is a map, dividing this world into two distinct territories. Outside of it are all the numbers that have the freedom of infinity. Inside it are numbers that are prisoners, trapped forever within this circle."

Ian Stewart: "If we think of a computer screen and look at each individual little element, each pixel of the screen, and pick one of these pixels then apply the rule lots and lots of times, either the pixel moves off and disappears completely from view or it stays forever inside a circle of radius 2, perhaps converging to a fixed point, perhaps falling into a repeating pattern or maybe never repeating."

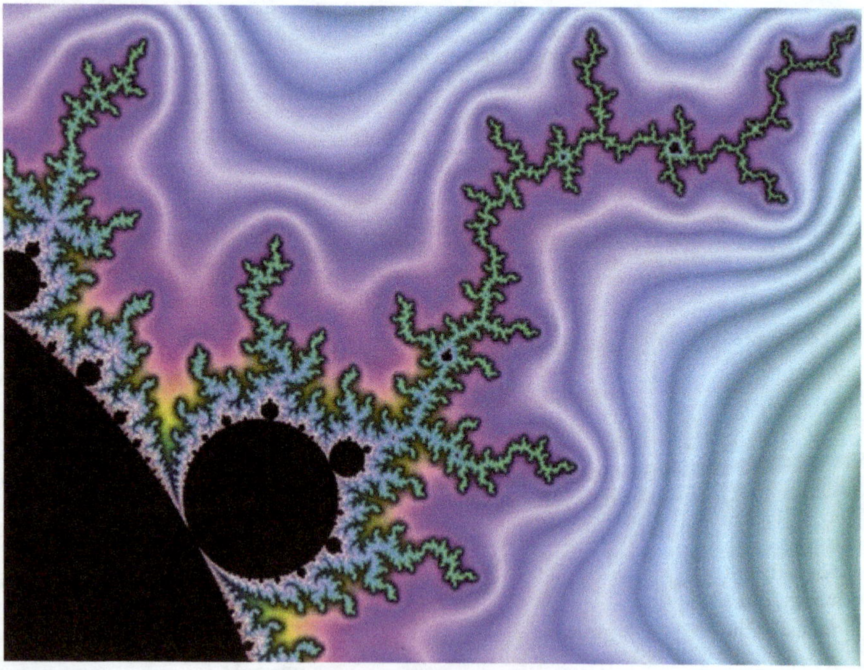

Arthur C. Clarke: "Now the colours are completely arbitrary. They could be anything, but they are not meaningless. The point of this process is to distinguish between going off out to infinity or going into zero. Any point that does not move out to infinity when the rule is applied is coloured black and any point that goes off to infinity is coloured in all sorts of wonderful rainbow hues based on how fast that point goes away. The important bit is the black bit in the middle. That's all the stuff that doesn't escape when the rule is applied. A very good analogy is the contour map of a mountain range, for example, where the contours are drawn and coloured. The highest areas might be coloured white, then brown, then green, and then out into the sea in deeper and deeper blues to show where the various levels occur. It is the same with the Mandelbrot set. The colours can be anything the operator chooses, but they do define the different areas of calculation. They can of course be changed at will, creating the most gorgeous results!"

Benoît: "When we get to very, very fine details of the set, the variations can become of overwhelming complexity. So complex in fact that no single picture could possibly do justice to them. It's impossible in black and white or even in colour to show how complicated it is".

Arthur C. Clarke: "No matter how much we magnify the set: a million times, a billion times until the original set is bigger than the entire universe, we will still see new patterns, new images emerging because the frontier of the Mandelbrot set is infinitely complex. When I say infinitely, I really mean that. Most people when they say infinitely, they mean 'oh, only rather big.' But, this really is infinity!"

No one knows whether the set is locally connected. Answering a question raised by Benoît, John Hubbard and Adrien Douady proved that the Mandelbrot set is connected and that you can get from any point in the set to any other point of the set without leaving it. Hubbard and Douady conjectured that the Mandelbrot set also is locally connected. If you draw a circle around any point in the set the part of the set inside that circle is still connected. Despite hard work by brilliant mathematicians, no one actually knows whether the set is locally connected.

No one knows either the exact area of the Mandelbrot set, though it is known to be at least 1.506 and not more than 1.507. Massive computational power has been expended on this problem, though it could take a while longer before anyone actually finds a use for the result! Maybe the boundary of the M-set has positive area. Benoit conjectured and Mitsuhiro Shishikura proved that the boundary of the set is two-dimensional.

Nevertheless, the fact that the Mandelbrot set exists and that it has an area is enough for mathematicians of the Everest school to attempt the challenge — because it's there. In the absence of any better ideas, they continue to chop the M-set up into ever smaller boxes, counting them all up to arrive at ever closer estimates. Convergence is very slow and it has been suggested that maybe the boundary of the M-set has positive area though no one knows the answer to this either.

Ian Stewart: "One of the most striking facets of the Mandelbrot set is the internal consistency of the object. It all hangs together. If we look at the boundary and zoom in and then look in just the right place, what we see are baby Mandelbrot sets perfect in every

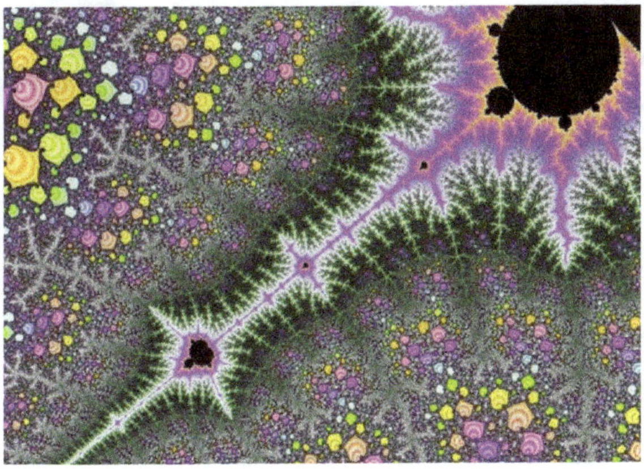

detail. They are in fact just slightly bent compared to the full image of the original cardioid, but that is very difficult to see. And they are decorated by slightly different external features. And then, by the same token, if we zoom into the boundary of those cardioids we see baby, baby Mandelbrot sets — the second generation. And inside those, we see the third generation. It goes on forever. And so, what we are seeing are islands of order in a sea of chaos."

Ian Stewart first saw the Mandelbrot set in the 1980s. He remembers it clearly. "We were at a mathematical conference on something totally different. And everyone went along to an exhibition because it was of mathematical pictures. And there were these amazing coloured pictures on the wall. I'd not really seen anything like them before.

The Mandelbrot set is the most famous fractal of all. It's not easy to describe it visually. It looks like a man. It looks like a cat. It looks like a cactus. It looks like a cockroach. It's got little bits and pieces that remind us of almost anything we can see out in the real world, particularly living things. It has a character that reminds us of a lot of things and yet it itself is unique."

There is indeed an infinite variety in the Mandelbrot set just indeed as there is in the world of nature. We see shapes that remind us of elephants' trunks, tentacles of octopae, sea horses, compound insect eyes.

Michael Barnsley: "The Mandelbrot set is real. An absolute thing — no question whatsoever! Any mathematician, any computer scientist or student in a school can study it and find the same thing. It's a common experience. And so, such things that can be magnified forever and have infinite precision do exist. But, they're not touchable."

Ian Stewart: "It's a geometrical shape, an icon if you wish, which somebody recently called *The Thumbprint of God*!"

The Mandelbrot set is one of the few discoveries of modern mathematics to be assimilated by society as a whole. It has appeared on mugs, T-shirts, record sleeves, and in pop videos even in cinema and television commercials.

Right: **Nigel with Arthur**

Far right: **Arthur in his garden with dogs**

Arthur C. Clarke: "Why I think there may be some connection between the Mandelbrot set and the wiring of the brain is because when I close my eyes and press my fingers against my eyelids I see these patterns. I'm sure you're all familiar with them. You also see them when anybody gives you a bang on the head! Sometimes these patterns echo some of the shapes of the Mandelbrot set. Also, I am told, though I have never tried this experiment myself that when certain chemicals are ingested you experience visual hallucinations strikingly similar to some of the patterns in the Mandelbrot set.

So why do these strange patterns have such an appeal? Well, obviously they trigger some kind of resonance in the mind. In Ecclesiastical designs such as stained-glass windows, particularly in Islamic art, we find many echoes of the Mandelbrot set centuries before it was discovered!"

Stained-glass windows

The M-set

When Benoît was exploring this set he never felt that he had invented it. He never felt that his imagination was rich enough to conjure up all those extraordinary things. He knew he was making a discovery. "They were simply there, even though nobody had ever seen them before. It's marvellous," he said, "that such a very simple formula creates all these very complicated things! So, the goal of science is starting with mess to explain by simple formulas. It's the kind of dream of science. And in this case the dream was implemented in a fantastic fashion."

The pictures of the Mandelbrot set are patently very beautiful and elegant, but what is their practical value? Arthur C. Clarke's response to this question was, "I'm tempted to answer in the famous words of Faraday who once said when someone asked him what use were these experiments — playing with wires and magnets — "What use is a newborn baby?" Faraday is also supposed to have told the Prime Minister, "One day, Mr. Prime Minister, you'll be able to tax it!" And in fact, fractal geometry has enormous commercial value. Fractal technology is now employed in the manufacture of capacitators, microwave invisibility cloaks, chemical mixers, military camouflage and soundproofing.

© William Rood

Arthur C. Clarke: "Often when I am looking at my computer screen and watching the beautiful images unfold I am reminded of Keat's famous lines in Ode to a Nightingale:

Charmed, magic casements,
opening on the foam of perilous seas
in faery lands forlorn....."

Heinz-Otto Peitgen: "There are many things that can be implied by the Mandelbrot Set. The phenomenon which strikes most of us as the most unusual about it is the fact that the formula, which is as short and simple as a mathematical formula can get, fully describes the Mandelbrot Set.

So we have this very simple, basic formula on one side and on the other side an object which, as it is visualised reveals an infinity, which goes much further than most of us could imagine possible. It is a structure which has infinite detail. If we look at one detail, within that detail we find an infinity of other details."

Heinz-Otto Peitgen

Though the stunning beauty of the images the Mandelbrot set generates appeal to us on many levels, the psychological reasons for this appeal are still a mystery. In his book *Aesthetics and Sociobiology*, the psychologist Daniel Berlyne proposes the idea that aesthetic appeal is a combination of novelty and familiarity. Without novelty, things get boring; without familiarity, things can remain incomprehensible. The Mandelbrot set is

surrounded by small copies of itself — there's the familiarity, but each copy is slightly different — there's the novelty. Perhaps there is some structure deep in the human mind that resonates to the patterns in the set. Like the M-set activity, life is richest on the boundaries. Between land and sea. Between earth and sky. Consciousness and life itself exist at the edge of chaos.

Arthur C. Clarke: "The Mandelbrot set is indeed one of the most astonishing discoveries in the entire history of mathematics. Who could have dreamed that such an incredibly simple equation could have generated images of literally infinite complexity and beauty? We've all read stories about maps that revealed the location of some hidden treasure. Well, in this case, the map is the treasure!"

06

THE FRACTAL GEOMETRY OF NATURE

Benoît published his revised and expanded version of *The Fractal Geometry of Nature* in 1982. Benoît: "After my book of 1975 — the small French book — my life accelerated very, very suddenly. First of all, the book was very well received and it was soon translated into English in 1977. For many years after that I travelled around not returning to Harvard. Suddenly, after this book of 1975, I was asked to be a visiting professor of mathematics. Given that in 1945 I had decided to forsake academic mathematics, that was a very great surprise to me. The year 1979-1980 was a very important one in my life — one of the high points again. I had an assistant, who was very good, and I asked him to make demonstrations in the class about using the computer in mathematics. The students were flabbergasted. It was something astonishing. They could not believe that one could look at pictures and then understand many things in mathematics which they thought were purely theoretical. But more important yet was the influence on my colleagues. First of all, they were sceptical: "Oh well, nice pictures," they said, "but…" Then when I was discovering more and more about the structure of what became to be known later as the Mandelbrot set, they were simply stopping me in the hall and saying, "May I see one of these pictures? What did you find?" An immense, fantastic wall — like the Wall of China — was just tumbling down into pieces before me.

Mathematicians had convinced themselves of this separation between the eye and the mind — the idea that mathematics could never gain anything from seeing. This was long past the time when pictures could help mathematicians and they were seeing in front of their eyes new discoveries being made by the application of the very patient analysis of many, many pictures."

Nature finds the same solution to many different problems. Like how to drain water from the land into the oceans, and how to get blood from our hearts to our fingertips and back again. The templates that nature uses are fractal. Clouds look the same at all scales. It is impossible to determine the size of a cloud from a photograph of it.

Clouds have now been shown to have the same dimension over 10 orders of magnitude, making them the most uniform fractal objects on the planet. Why should this be? Clouds are formed by the condensation of tiny droplets of water which occurs on a fairly random

basis under suitable conditions. However, once these are formed, they tend to attract more tiny droplets at certain points around them. This creates the necessary conditions for a fractal.

After Benoît returned to IBM in 1980, he continued revising his 1977 book, which he felt had become obsolete even though it was only seven years old. He wanted it to include some of the newest discoveries in both mathematics and physics. Benoît: "*The Fractal Geometry of Nature* is the best book I have written — the big book. The others were much more specialised."

In the summer before that book appeared, a meeting was held in the French Alps in Courchevel in 1982 sponsored by IBM Europe. The Physicist, Professor Bernard Sapoval, met Benoît for the first time at that Courchevel conference in Switzerland. Sapoval recalls that what was really important about that meeting was that there was a multi-disciplinary, pluri-disciplinary audience there. "For me that was new because I had been raised and made my career in Solid State Physics, working in the highly specialised domain of semiconductors. But here there was a mixed audience of mathematicians, statisticians and physicists."

All fractal conferences always turned out to be multi-disciplinary. The physicists applying fractal geometry were working both in chemistry and in geology. Carl Evertsz: "If you went to a conference like that you would certainly learn a lot of science!"

Bernard Sapoval was one of many physicists who were trying to find out why the structures in the human body like the bronchial structure are fractal. The question, which is still wide open, is why does nature choose to apply fractal geometry and not another geometry in the human body?

Benoît at a fractal conference in the 1990s

Benoît with his book

Benoît: "This idea which had been bottled up very tightly in the book of 1975 and was not very well-explained in the English translation of 1977 became very much clearer in the 1982 version. Now, of course, it has become everybody's passion. The book of course had many pictures."

Computers had improved in the meantime and one person who worked hard with them was a close colleague of Benoît called Richard Voss. He came to IBM as a physicist, but turned out to be about the best programmer of the time — in particular with computer graphics. He devised systems that were not really supposed to function on those machines. They were designed for other work. Voss put together a system which yielded pictures of extraordinary beauty. He was the only man at that time who was able to do this. He made mountains that never were and clouds that never were. He made imitations of nature which were so lifelike that many colleagues refused to believe that they were not paintings or perhaps, even, photographs.

Benoît's book of 1982 changed a large number of people's view of what science could achieve; how difficult and advanced mathematics could be combined with very hard problems concerned with 'messy phenomena' and come out on top.

Benoît worked in many different and varied fields. He was a man who was extremely open to adventure and, indeed, to anything new. When he came to Yale, Benoît had two people working for him, who were computer scientists. They were working on rendering 3-D objects. They were generating forgeries of landscapes.

Computer-generated fractal landscapes

He was working with computer graphics designers using computers in a very strong fashion. They were artists. They were using fractal geometry to make these forgeries of landscapes, making them beautiful and to look as realistic as possible.

After the publication of his *Fractal Geometry of Nature* in 1982, it was jumped on primarily by the physics community. Benoît had a lot to do with physicists. He had a lot of friends who were physicists so it was natural for him to work with them. Physicists could see that this new geometry of nature was something which was obviously going to be extremely useful to them. They looked through the book and were immediately motivated. They wanted to try out the applications themselves. They asked, "Why are all these structures fractal and how do they arise?"

Computer-generated fractal landscapes

Benoît: "Almost every page of my book could have been published separately as an article. I kept it to 500 pages because I believed that this would keep it under control! Sometimes, however, I was accused of being too concise. And many people who read my book said, "That could not be true. He's just making it up. So many of them did in fact set out to prove me wrong. Immense literature came out from people who tried to prove me wrong, but who eventually proved me right — much more right than I ever had any hope of being! To put it most simply, it was the right idea to have at the right time and I was fortuitously in just the right place."

In mathematics, ideas which are elegant, simple and very difficult are the ones which are the most highly esteemed. Benoît's contribution to mathematics has always been through asking questions which others did not ask because they did not have his kind of mind and approach to scientific issues.

Benoît: "Since 1982, I have done many, many things. I have continued to work in physics and I have certainly done new work in mathematics. I have come back to the issues of financial prices with a vengeance. I would not say that things have become completely organised. They have not. It is a matter of finding new evidence for the richness of the same set of ideas rather than changing the whole picture completely. Fractal geometry was by no means finished in 1982, but it did take shape in that year."

© Gabriel Lesmoir-Gordon

07

THE ART OF FRACTALS

In Painting, Architecture and Music

Looking back, Benoît reflects: "I had this experience which many people repeated and told me about: that when I first saw images of the Mandelbrot set I was the first person to see them! There was absolutely no way anybody could have seen them before. Yet after a few days, or sometimes a few hours, a few minutes, they became almost familiar. I was finding features in them which I had seen somewhere. So, where have I seen them? Well, first of all certainly in natural phenomena but also in art. One wonders, why is it so? We know the brain has some cells which handle shapes, boundaries, and other cells which handle colour. Does the brain also have cells which handle fractal complication? We don't know. It's a purely hypothetical question. It's a tempting question, but we don't know anything about it. Some of my ideas had certainly been around for a hundred years which is really very recent because they are in fact the most ancient ideas imaginable. This brings me to the most attractive and surprising aspects of my work — its position in a certain magic circle. Fractal geometry did not begin with me. That's for sure. I organised it. I conceived of it. I developed it. As I've said, some of the ideas were at least a hundred years old when I started. But I was completely wrong about the date because these ideas were actually millennia old. They permeate the decorations of Egyptian temples, of Persian temples, of Hindu temples and of European churches. They even correspond to descriptions in the books of Zen Buddhism. In general, man's imagination in religious iconography is filled with fractals. It's an interesting question for me whether the mathematicians of 1900, who defined those early proto-fractals were aware of the history of those objects. Probably not. But whatever the case, they certainly did not start from scratch. They defined a certain structure and I went on and developed that structure. This beauty was there in a certain sense all along — from the very beginning. There is a circle from beauty to mathematical abstract thought to science, to beauty again. It is a magic circle of the most pleasing shape."

In every particle of dust there are Buddhas without number

In the third century, Buddhist Avatamsaka Sutra Indra's Net is used to describe the interconnectedness of the universe. Far away in the heavenly abode of the great god Indra there is a wonderful net which has been hung by some skilled craftsman in such a manner that it stretches out infinitely in all directions. The craftsman has hung a single glittering jewel in each *eye* of the net, and since the net itself is infinite in dimension the jewels are infinite in number. In each one of these jewels are reflected all the other jewels in the net. Not only that, but each of the jewels reflected in this one jewel is also reflecting all the other jewels.

There is evidently a parallel to be found here in the infinite nature of Indra's Net and the infinite Mandelbrot set in that each one of the jewels reflects all the other jewels in the net. As Ian Stewart says of the M-set, "…if

Ely Cathedral

we zoom into the boundary of those cardioids we see baby, baby Mandelbrot sets, the second generation. And inside those we see the third generation. It goes on forever."

Art and buildings of different periods show examples of both Euclidian and fractal geometry. Tower blocks, composed of simple pieces all about the same size, are Euclidean.

By contrast, Ely Cathedral with its profusion of spires and turrets shows the repetition of the same patterns over several scales, suggesting a fractal interpretation.

This structure is a common feature of Hindu temple design too. The Kandariya Temple in Khajuraho in Central India resembles a mountain peak, composed of a cluster of a cluster of smaller peaks each made of a cluster of still smaller peaks.

Fractal structure was forcibly replaced by the cereal-box or barracks style, but it is making a comeback. Fractals can be found in the architecture of some African villages too, such as the Ba-ila settlement in southern Zambia. The settlement is a ring of rings. The entrance is on the left. Arranged around the settlement walls are smaller rings, the dwellings of extended families. A social gradient is reflected in distance from the entrance, so the dwellings become larger as we move further from the entrance. Each extended family dwelling has a pattern

The Kandariya Temple in Khajuraho, Central India

similar to the whole settlement.

We may expect fractals to be found more often in nature, and Euclidean shapes more often in manufacture, but fractal designs have always appeared in architecture whether for motives of utility, aesthetics or the desire to mimic nature. In painting too we see different geometries. Beauty is in the eye of the beholder, and different people do see the world differently. For example, Cubist painting shows fractured planes, uncertain volumes and shifting, multiple viewpoints. Its aim was to replace the single time and space of the renaissance perspective. Other artists focused on the body, viewed from the outside. The surface of the skin is mostly smooth. The body hides its fractals within!

In the 1830s and 1840s the Japanese artist Katsushika Hokusai produced marvellous woodblock prints which clearly captured fractal aspects of nature. The woodblock *In the Hollow of a Wave Off the Coast of Kanagawa* from his *Thirty-Six Views of Mount Fuji* is familiar to many.

Over his long life Hokusai observed that nature shows patterns that were then nameless, but which we now call self-similar, and these are frequently reflected in his work. Immediately after his *Thirty Six Views,* Hokusai drew *One Hundred Views of Mount Fuji* in which his grasp of the fractal aspects of nature achieved a sophistication rarely rivalled even today.

Katsushika Hokusai. In the Hollow of a Wave

Katsushika Hokusai. One Hundred Views of Mount Fuji

© The British Museum

Max Ernst. Europe After the Rain

© Wadsworth Atheneum Museum of Art

Like a number of other 20th century artists, the surrealist painter Max Ernst used decalcomania. Ernst used the visually suggestive textures of paint in *Europe After the Rain* and several other late paintings.

Fractal notions are found even in ancient Egypt. The capital decorations of this column from an Egyptian temple represent a lotus blossom, symbolising the many scales of the development of the universe. We can recognise it as successive steps in the development of the Cantor set.

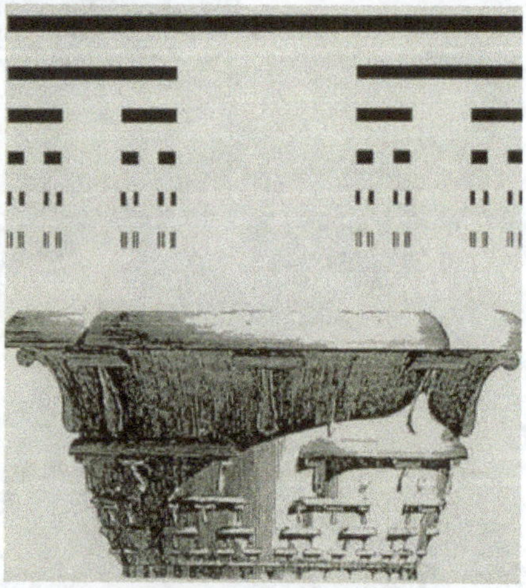

The Cantor set revealed in a Greek temple column

Above:
Salvador Dali. Visage of War. 1952 © Salvador Dali, Fundació Gala-Salvador Dalí, DACS 2017

Salvador Dali's *Visage of War* represents the infinite horror of the Spanish Civil War during the 1930s. Note the repeating pattern of eyes and mouths containing skulls whose eyes and mouths contain skulls and so on.

Interestingly, Dali's preliminary study for the painting did not show this. He discovered self-similarity as a more powerful way to express unending horror through the implied infinite regress of copies within copies

Jackson Pollock Totem Lesson 2, 1943 © The Pollock-Krasner Foundation ARS, NY and DACS, London 2017

Jackson Pollock increased the complexity of his patterns as he continued to experiment with the drip technique. Analysis suggests that

Above:
Jackson Pollock Blue Poles Number 11. 1952 © The Pollock-Krasner Foundation ARS, NY and DACS, London 2017

Right:
M-set with Escher Lizards

© William Rood

Pollock refined his technique over the years. The fractal dimension increases from about 1, sparse and simple in 1943 to 1.72, thickly layered and complex in 1952.

While the mechanics of paint falling from the brush contribute some of the observed structure, much derives from Pollock's movement across the canvas. When asked if he painted from nature Pollock replied, "I am nature."

Benoît was struck by the recursive geometry of the Dutch conceptual artist Maritus Escher (1898–1972), the undisputed master of tessellated art: "For a long time I have been interested in patterns with motifs, which get smaller and smaller until they reach the limit of infinite smallness." It turned out that Escher's strange reptiles can be mapped onto the exterior of quadratic fractals allowing the creation of tessellation with fractal limits.

In 1954, the young mathematician Roger Penrose visited an Escher exhibition with his father. Intrigued by this almost paradoxical use of space and form, Roger went home and designed the impossible tri-bar while his father created an apparently endlessly ascending staircase. They published their creations in a psychology paper. Escher went on to use both these ideas in subsequent prints. 1979 saw the publication of an extraordinary book by Douglas Hofstadter, cross-pollinating ideas from mathematics, psychology and art.

Hailed at once as a masterpiece, the book went by the name of *Godel, Escher, Bach: An Eternal Golden Braid*, and on its cover was the impossible tri-bar. While Godel's celebrated theorems had been hailed by philosophers as proof that humans are fundamentally different from machines, Hofstadter argued that Godel's results and those of Alan Turing are just as limiting for humans as they are for computers.

Fractals are clearly one avenue through which art can express some of the rich patterns of nature. Benoît: "One of the main joys of my life is music. I learned music but very late because of poverty and as a result of moving from one place to another and because of war events. When I was about 20, I was taken by a friend to a symphony concert. After a few such outings I was amazed that I was like a sponge, just absorbing everything as if I had been prepared forever for this kind of activity. I became very, very fond of music. And, in a certain sense, the more complicated the better. The music I love most is opera, which combines action, theatre, voice, instruments — everything together. The story is often very silly, but we do not necessarily listen to the story. The relationships between the visuals are very often deeply profound and I find it an ever-renewing pleasure to witness the same great operas, which I know by heart, again and again. I also enjoy painting but not quite so intensely as music. We have a small collection of paintings at home — mostly by our friends — and I go to exhibitions very

often. I think that music is more bottomless for me than painting has ever been."

Music has fractal structure. This is an example of where the fractal principal is essential: any piece of music that does not have structure on many scales tends to sound boring after a while and the same goes for paintings too.

Benoît: "When my book speaks of the fractal geometry of nature, the word *nature* is to be taken in the very broadest sense. The structures which I study include those which are clearly man-made, but very complicated and so big that the design of the whole is not precise and there is much looseness about it. This includes economics and linguistics as well as music. Music is certainly a human activity, a creation of man. There's been an immense amount of study about the specific notes, their intonations and small structures as well as big structures like themes and repetitions. Over the years two very great composers, Charles Wuorinen in the United States and György Ligeti in Europe, have both come to me with the same story. They both said that, even though they were very highly-trained classical composers and were aware of all the techniques of composition, they had the impression that something was missing in their understanding of what music was. In particular on the central issue: what distinguishes music — good or bad — from plain, old noise? Sometimes the distinction is in dispute. Both concluded after reading my book that music must have a fractal structure. Wuorinen as well as Ligeti felt that this gap in their understanding of their craft was filled only by this geometric notion."

Charles Wuorinen: "As a result of my encounter with fractal geometry, I had a deepening sense of the significance of organisation as opposed to the objects which are organised. In a very real sense one can't really make a distinction. It's like the distinction between form and content which is basically nonsense. A form is not a bin into which one throws a content, certainly not, but still if one has all these notes to push around in various places to make a composition one can often think of them as objects and often come to believe in their individual special properties.

Yet a more *fractally* oriented view would look at them or listen to them not as objects but as locations in a kind of space/time — a kind of pitch space and time space in which the relations between them, the intervals of time and the intervals of pitch that concatenate an immense complexity to make a given composition really matter."

Benoît: "So music is fractal. Does it mean that you can use algorithms to compose music? Well, you can, as a matter of fact. It's been done by many, many people. Much of it is perceived by listeners as being bad music. Bad music is a high compliment. It's much better to make bad music than to make plain old noise. Richard Voss has used

Charles Wuorinen

defective transistors, which create noises, to create so-called music. And it's always perceived as being bad or incomplete or somewhat unprofessional music. But again, that's a very high compliment. So many of the works of man of the most obvious and the least obvious kinds do have the same structure as the works of nature. I would add that large computers have fractal structure. The internet certainly has fractal structure. Fractal geometry has helped us to understand the internet and to improve it. It was found to be a fractal — a multi-fractal — and with very high precision. The essential thing to know about the internet is how it functions. Computer memories are fractal also of course."

Charles Wuorinen: "Many processes and objects in the world are fractal in nature. We have discovered that music — not just my own method — but music as a whole is intrinsically fractal unlike the visual arts in which particular representatives can be made fractal but in which the entire field is not. To discover this about music is to answer a question that has agitated musical minds since antiquity: "Why do we feel such a strong visceral connection between music and the natural world?" Classically this has always been described in very naïve terms musically by reference to the partial series and other related matters and this goes back to Pythagoras. I think that the real answer lies in the fact that music is, as I've said, intrinsically fractal, as is a very large part of the natural world. This is the connection."

The emerging crop circle art phenomena has surprisingly also produced some remarkable and intriguing mathematical works. In a field close to a remote East Anglian village called Ickleton near Cambridge on the night of 11 August 1991, an enormous and astonishingly detailed representation of the Mandelbrot set made its appearance. It was not visible from any road, but was spotted by the pilot of a light aircraft flying from London to Newmarket. 'Cereal Researcher' Beth Davis reported the event to Benoît at the time and she recalls: "When Benoît Mandelbrot was told of the formation, he said he was very gratified to hear of the theory taking root!"

Benoît: "It's certainly pleasing to be remembered in this way. But I can tell you, I plead not guilty. Was it a student joke? I don't think it's the work of extra-terrestrials, I can't wait to see what the next one will look like."

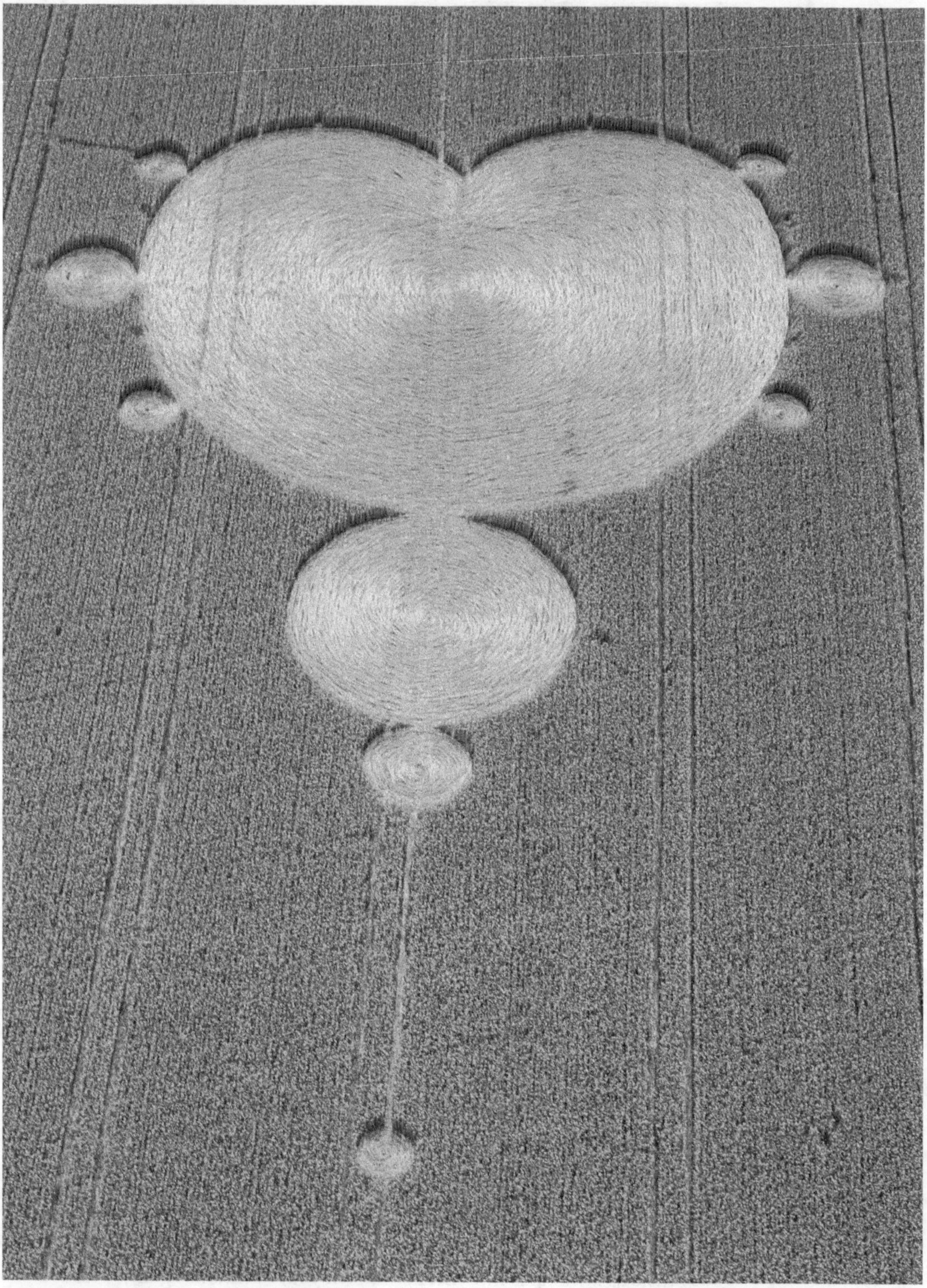

The Ickleton Mandelbrot set Crop Circle

George Wingfield, a retired Royal Greenwich Observatory and IBM engineer was one of the first to visit the site after the discovery of the formation. The quality of the craftsmanship immediately made a big impression on him: 'The sheer perfection of the lay of the crop was quite apparent to all those who actually saw it at ground level. The wheat was swept flat with a flow which showed no trace of the irregular jerky movements so often apparent in other circles. The tiny circles tangential to the perimeter of the great cardioid were minute works of art in themselves: neat unbroken swirls of scarcely 18 inches in diameter touched the main body of the Mandelbrot set with confident precision and immaculate circularity.'

The debate was opened up. Letters filled the pages of the Cambridge papers and rumours abounded. The land adjacent to the field where the formation was made was farmed by Dick and Mary Wombwell. This is how the farmer described his first visit to the site: "It was overwhelming. The wheat was laid down in a clockwise manner. Each stem was bent over at the same height above the ground — about half an inch. The stems were not crushed, and the ears of wheat were intact. The seed was still in the husk even though the crop was about ten days from harvesting. No one could have made this in the dark and if they used lights to aid them we would have noticed them from the farmhouse."

Montague Keen an agronomist and research member of the Centre for Crop Circle Studies was assured by Cambridge University's Vice-Chancellor that no action would be taken against any student who confessed. Keen opened up the issue in the Varsity newspaper. This magazine circulates to about 10,000 students. Keen appealed for the hoaxers to come forward. Just one letter followed this appeal. The writer, who said that he was the mathematician suspected of the hoax, declared once and for all that he had not done it. Then another rumour surfaced. Arthur C. Clarke announced that he knew who had made the Ickleton Mandelbrot. Clarke had even named the culprit, a Cambridge mathematician called Cade Roux. This mathematician also rapidly denied the rumour! The mystery remains unsolved.

08

PRACTICAL FRACTALS

John Archibald Wheeler, protégé of Niels Bohr, friend of Albert Einstein and mentor of Richard Feynman, wrote:

"No one will be considered scientifically literate tomorrow, who is not familiar with fractals."

Fractal growth methods are now used to model marine organisms such as sponges and coral. Scaling is also of interest in case studies of vegetative ecosystems, earthquake data[a] and the behaviour of density-dependent populations.[b] Fractals also hold a promise for building better roads and for designing ships that are less likely to capsize.[c]

Another interesting geometrical application is percolation which is the process of a liquid slowly passing through a filter. It's how coffee is usually made. A critical value in a system determines the percolation threshold. The percolation threshold determines why some forest fires die out while others continue to spread. As the temperature soars in a plantation of evenly spaced trees on a very hot dry day, the odd leaf or twig ignites sending a whole tree up in flames. This is an essentially random process — the factors involved are beyond our powers of prediction. But once a tree is in flames, the fire easily spreads to neighbouring trees and this process can now be modelled with iterative techniques. So we can't say anything about which tree will go up first, but we can say how the fire will spread. The fire will have a fractal boundary.[d]

This principle applies as much to the spread of infectious diseases or magnetic polarity as to forest fires. It is essentially the same process unfolding. What we find is a highly sensitive dependence on initial conditions, producing phase transitions. Once a critical threshold is passed, the fire spreads outwards, the disease becomes an epidemic, while the material becomes magnetic. This threshold is also important in determining when an epidemic may become a pandemic and when iron becomes magnetic. Fractals are related to the percolation threshold because near the percolation threshold self-similarity reigns supreme.

Ian Stewart: "I think the discovery of the Mandelbrot set and of fractals in general is very important. There are some applications, but it's not yet put an important new gadget into every home whereas things like the silicon chip certainly have. So most mathematical developments are like this. The ideas must come first, and then you have to translate them into practical things."

Michael Barnsley: "No longer do you have to draw a straight line through your data to make science of it. You can actually draw some fractal curve through it or measure some fractal dimension of the data and do science. So the first application is in terms of a better description of the physical observable world. There's a new branch of mathematics now available to all scientists and that application will stretch on through the centuries. The primary application will be as a tool for

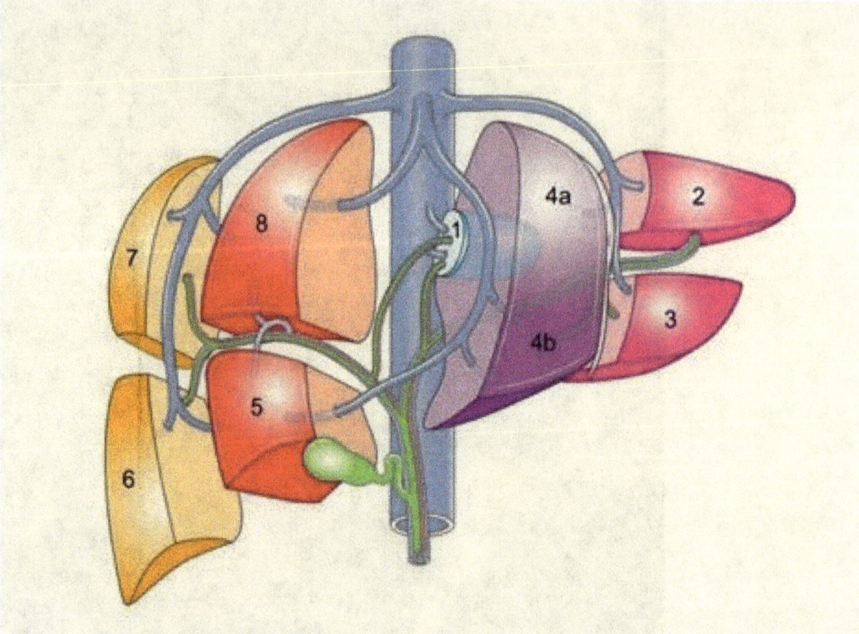

science in its own right. Science, and then engineering and on through into the building of the next generation of devices and equipment which will follow from that in terms of the sort of application that we think of. There may be a new type of — not computer — because before we perceived and understood about desk-top and lap-top computers they were not here. We didn't imagine them. But there will be new devices, extraordinary new devices based on the principles of fractal geometry which will emerge over the next century."

Carl Everstz: "If a surgeon looks at the liver of a patient he is going to operate on, he finds that the liver is smooth on the outside, but that inside it is permeated by fractal trees. There is the portal vein. There is the liver artery. There is the bile tract and the liver vein. There are four intertwined trees in the liver. Clearly wherever a surgeon cuts he will be cutting through blood vessels.

Looking at the portal vein it becomes clear that it is a fractal tree, which subdivides the liver into various segments. Each segment is irrigated by one particular main branch of the portal vein. Importantly, the segments or domains have almost smooth boundaries. If there is a tumour inside the liver the surgeon would want to identify in which domain of irrigation this particular tumour lies so that when he operates he can take away just one segment. If just one branch is cut out of a tree the other branches are not damaged. It is the same with the liver."

One problem that radiologists encounter when they are trying to make images of the liver using CT or MRI scans is that they do not see the full extent of the portal vein because even the thickest branch has a diameter of just two centimetres and the smallest one has a diameter of less than one millimetre. The resolution of these machines is not high enough to capture the smallest branches. When an image is made of the portal vein within the liver radiologists see a kind of castrated tree. It is pruned. They do not get the full resolution. This is where using fractal extrapolation can predict where the end points of the portal vein are found and therefore where the surfaces between the various domains of irrigations are located.

Carl Everstz: "So now we have a method that uses fractal geometry. When a patient comes in and we get a CT scan of the incomplete portal tree, we use methods based on various aspects of fractal geometry to compute from this pruned version of the portal vein a complete version. This gives us the boundaries of the various segments in the liver. It's then possible to identify in which of these domains of irrigation the tumour lies. Through this three-dimensional visualisation we can tell the surgeon before the operation where the tumour lies and where the segment boundaries lie so that he can precisely plan the surgical procedure."

Ivar Giaver has used Benoît's geometry in one very important aspect of his work in biology. He works with cells in tissue culture and he measures these cells electrically. When he does that he creates a time series. Benoît's work is very adaptable and useful in the analysis of these time series. What Benoît did in this regard was in fact to generalise Brownian Motion.

Michael Frame: "By training, I am a topologist and I saw immediately that there were some real possibilities for using fractals to study some problems in topology. In a sense, the last 15 years had been a detour for me, learning to use these tools, but the tools are so incredibly fascinating that I haven't actually got back to the topology yet! I hope I will sometime, but…"

Benoît: "Somehow fractals are seen to correspond to something in our mental set-up which people find friendly. Fractals are not harsh or hostile. They are indeed friendly. We know them. Most of my friends, who started on complicated courses, have almost invariably ended up teaching only the fractals to the mathematicians. This friendliness is a great mystery, but it's a mystery which improves upon deepening."

Benoît found his most enthusiastic acceptance first among applied scientists working with oil, rock or metals. Several specialists in oil research have started work on fractal problems. Fractals have also become an organising principle amongst physicist in the study of polymers.

Benoît: "Phenomena of great irregularity are obviously very wide-spread throughout nature. In the study of Condensed Matter — Polymers — of such physical problems, we find shapes of extremely great complication. These shapes couldn't be examined as geometric shapes before the development of fractal geometry because there was no language to describe them. We could only say things indirectly about them — If we make such and such a measurement we will get such and such a result. But that was, in a certain sense, a shadow of the object. It was the effect the object had on a certain measurement procedure, but the object itself could not be described with regular geometry. This is a very mundane example which is just the tip of the iceberg. We now know that there are an enormous number of structures which are only describable with fractal geometry."

Butter is made up of chains of saturated fats and this arrangement of fats can give rise to beautiful fractal structures if prepared in the right way. If a small quantity of butter is sandwiched between two glass plates and one of the plates is pulled up air invades the sticky butter. The fat chains hang together in a way that forms these beautiful branches.

As batter heats, bubbles of many different sizes are formed producing beautiful fractal ring and spot patterns not unlike craters on the moon which were of course formed by entirely different processes. Bread is fractal. On the outside bread looks smooth and Euclidean, but when it is sliced open its fractal nature becomes apparent. It is made up of a foam of many-sized bubbles baked to a solid illustrating the point that with many fractals what is left out is just as important to the structure as what is left in.

Bread is highly structured, like sponge. The same structure recurs on deeper and deeper levels.

The key to making bread is yeast. Before it is baked, the dough rises because yeast converts starch into carbon dioxide forming bubbles. Wheat contains gluten proteins that stretch out forming membranes like balloons which hold the bubbles in place. Baking turns these membranes into the solid making up the bread. Kneading the dough breaks up the larger bubbles. The more bread is kneaded, the more uniform the size of the holes and the less fractal is the bread. Mass-produced bread is kneaded

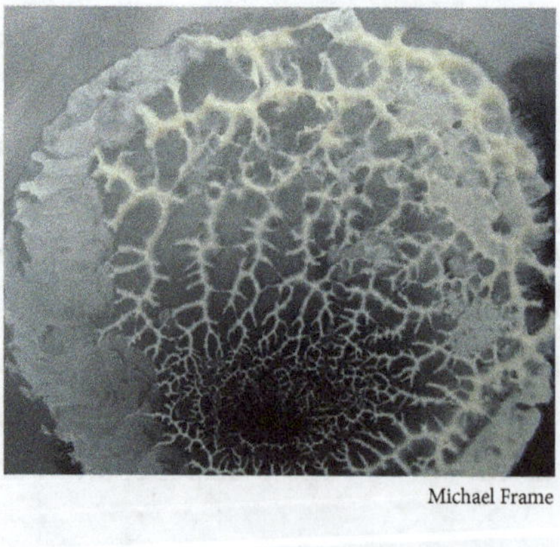

Michael Frame

Michael Frame

Right:
A butter fractal

Far right:
A pancake fractal

Right: **A white bread fractal**

Far right: **A rye bread fractal**

Right: **Diffuse limited aggregation in nature**

Far right: **Computer generated DLA image**

a lot so all the holes are small and are all about the same size. It is not fractal.

In the classic Hele-Shaw experiment, water is fed through a small hole in the middle of a flat layer of oil sandwiched between two sheets of glass. At first the water spreads out uniformly, but fairly soon the increasing straight interface between the fluids becomes unstable and breaks into fjord-like fronds, which fan outwards in a fractal manner reminiscent of coral or plant growth.

Viscous fingering strongly resembles diffuse limited aggregation. Both have a fractal dimension around 1.7 and there is growing evidence that they are mathematically related.

Coral grows outward by aggregation. That is, successive deposits of material attach to an outward-growing surface. They look very much like trees which grow outwards by branching from the inside. This is similar to snowflake forming and to DLA patterns.

Fractal geometry is basically an extension of classical geometry, an extension which allows us to consider, measure, explore and model forms of nature which we simply could not do before the advent of fractal geometry.

It is indeed an entirely new way of looking at nature. It is a new way of dealing precisely and mathematically with the patterns, the forms and the structures which appear in nature.

Dawn over the Himalayas

Fractal geometry has brought the eye back into science with a re-emphasis on pictorial images.

Fractal geometry is helping increasingly in the detection of cancerous cells *in vitro* and in describing precisely the progress of oral cancer. The surface structures of cancer cells are fractal at various stages of their growth development.[e]

Fractal geometry is providing faster and more accurate methods of interpreting mammograms. Fractal geometry also explains the unusually long incubation period of the AIDS virus.[f] The dynamics of the AIDS virus in the body starts to go chaotic as the immune system begins to fall apart. We are now beginning to understand why the AIDS virus has such a long incubation period, why finally after 10 years or so of being HIV positive, it kicks in and AIDS develops.

An astonishing explanation for some of the mysterious power law patterns found in nature like the three-quarter power law have been found using fractal modelling techniques. Physiologists have an empirical understanding now of how blood flows through the circulatory system and how this relates to the actual size of the vessels. Analysing all the data a pattern emerges and it is a three-quarter power law: the way one thing varies with another, based on the cube of the fourth root. Many of these three-quarter power laws have emerged from the measurement of unrelated systems. This is a powerful application and a highly beneficial aspect of fractal geometry.

Michael Barnsley's team at Iterated Systems in Atlanta developed a system to use fractal geometry for video compression. Michael Barnsley: "I woke up one morning and knew that I'd discovered it. This was the total secret to Fractal Image Compression: how to automatically look at a digital picture — these ones made at the low resolution input like our eyes receive — and how to turn it into a) a formula and b) an entity of infinite resolution. So, the goal was to be able to capture this Fire of Prometheus, this Fractal Wonder, put it in a box and then make it available to everyone."[g]

Barnsley has discovered how to make computers actually *see* images in a way that mimics the human brain. Our brains produce images of far higher resolution than the signals recorded by our eyes. Barnsley's mathematics mimics our minds by capturing the fundamental patterns of nature. His work has already resulted in a radical approach to interpreting and delivering images with computers. His mathematics could revolutionise how we store, retrieve and even how we fundamentally think about manipulating information electronically. Barnsley's discoveries in data compression technology cleverly define reality in mathematical terms, delivering pioneering theories and major practical applications.

PRACTICAL APPLICATIONS SOURCES

[a] *A Fractal Model of Earthquake Occurrence: Theory, Simulations and Comparisons with the Aftershock Data*
Pathikrit Bhattacharya, Bikas K Chakrabarti and Kamal.
Published under licence by IOP Publishing Ltd.
Journal of Physics: Conference Series, Volume 319, Number 1.

[b] *Introduction to Crowd Science*
G Keith Still.
CRC Press 2014.

[c] *Applications of Fractals and Chaos.*
Crilly, Earnshaw, Jones.
Springer-Verlag 1993.

[d] *Fire's Spread Looks Fractal.*
Philip Ball.
Nature (13th August 2001).

[e] *Fractal Patterns seen on Emerging Cancerous Cells.*
Hamish Johnson.
Physics World (March 2015).

[f] *Fractals in Biology and Medicine.*
Nonnenmacher, Losa, Weibel.
Springer 1994.

[g] *The Colours of Infinity.*
Nigel Lesmoir-Gordon, Arthur C Clarke, Benoît Mandelbrot, Ian Stewart, Michael Barnsley.
Springer-Verlag 2010.

09

IN RETROSPECT

The goal of science is to find the language which describes and explains the way the universe works. Geometry is the language scientists use to describe the physical structure of the space in which we live.

Carl Evertsz: “There were certainly people working on self-similarity at the end of the 19th century and at the beginning of the 20th century. There were philosophers and mathematicians discussing fractals even then. There are passages in the writings of Immanuel Kant where fractals and self-similarity are described in terms of the stars in the universe. Then take Hausdorff, take Cantor, take Sierpinski and Peano: these are mathematicians who dealt with self-similarity and who worked with it mathematically. Fractals are everywhere. Once you know them then you see them everywhere though no one but Benoît came through to develop the fractal geometry of nature.”

Benoît: “Looking back one of my greatest achievements — and it may very well have been my greatest achievement — was to bring the eye back into hard science after a very long absence and to show its endless creative power not only for one specific topic, not only for two specific topics, but in very many different and varying fields. I’m astonished by how little my predecessors used their eyes, how little they knew about presenting their evidence in a fashion which the eye could grasp. I hope now that my example will be followed and improved upon.”

Michael Frame: "The first time we taught the fractal class at Yale — the first time it was offered — it had no history at all. It was a brand new course. The people at Yale said, 'Oh well, you might get 20 or 30 students.' In fact, we had over 170. It was by far the largest math course in Yale's history! Benoît was ecstatic. The students weren't just sitting there being bored. They were engaged and they did really good work. This is the true measure of the size of this revolution: that you can get students who are not from a strong scientific background deeply interested in a science topic. I think this may be the biggest tribute to the power of Benoît's work."

The Yale mathematician Professor Peter Jones: "On any number of occasions in the past 30 years Benoît has been first obviously wrong, and then viewed as, well, sort of right, and then people would finally say, 'We knew that all along!' This has happened over and over again.

There was always a fierce onslaught by the defenders of whatever faith they had. In the end, they have essentially collapsed under the pressure of one single man who has enjoyed pushing his point of view and simply asking for people to examine the truth. The main reason that Benoît was 'the voice crying in the wilderness' for so long was that he was absolutely committed to the way things *looked*. He stayed that way even when it was unpopular to do so. Now through the force of computer graphics and the marvellous conjectures that

Peter Jones

this has opened up most people in the scientific community have come around to this way of thinking. Returning science and mathematics to the visual and tactile will be his most lasting legacy."

Benoît was always willing to do things which others looked down on and dismissed. He had faith in his own insights and for many years it was an unshared faith. But that did not disturb Benoît. It did not daunt him. It was not easy to stick to it and to persist, but he always had faith in his vision and that vision has subsequently been wholly justified. Many people might have taken a glancing look at these things and when they did not yield or when people said it was nonsensical they might have gone on to other things. But not Benoît. Benoît had vision, faith and persistence and it is those qualities that make all the difference.

Ivar Giaever: "There was a lot of controversy around Benoît because he was not really a mathematician. If you were a mathematician you did not like him. And there were a lot of physicists too who did not like him. I don't know why, because he has radically changed the way that scientists can now look upon the world. A lot of these things were known in bits and pieces, but he really put it all together and made it accessible to everyone. There are many people who think that Benoît should have received a Nobel Prize and I agree with that, but it's hard to find a field in which he should get it because there is no Nobel Prize for mathematics and fractals are really not physics in the strictest sense. But clearly he deserves to be recognised at the very highest level."

Heinz-Otto Peitgen: "Benoît was never part of mainstream science. He was always doing important work, but not as a mainstream mathematician or scientist. I believe you can only do that, support yourself, support your beliefs, go on and do it if you believe that you are doing something special. He always had the feeling that he was onto something hot, something very important. He kept the faith and eventually he managed to convince us all."

Ralph Gomory: "I think the most important thing is that those mathematical curiosities like the Cantor Set, the Sierpinski Gasket, the Peano Curve have turned out to be a pervasive part of our everyday lives. Like when you think about a simple notion like the length of the coast of Britain and you suddenly realise that that is quite meaningless.

Now these things that surround us like our lungs, the branching of the blood vessels, the branching of streams can be given a precise meaning. Fractal geometry has changed our view of the world. That's very impressive to me."

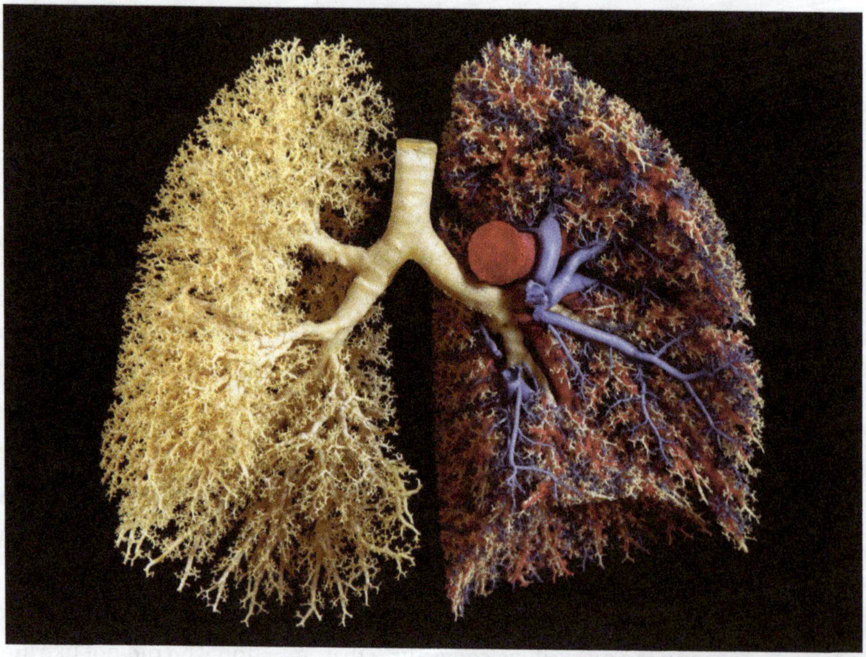

The Lungs

Peter Jones: "I think the most fascinating aspect of Benoît was his eternal curiosity and his willingness to look at things that everyone thought were well understood. Whenever he did this he discovered that what was considered to be well understood was either wrong or was misleading or was only a very small part of the story."

Benoît: "Looking back I have been astonished at how little my predecessors used their eyes, how little they knew

Richard Olsen

how to present the evidence in a fashion that the eye could grasp. I hope that my example will be followed and improved upon."

Economist Richard Olsen: "To me the greatest insight of Benoît's work is the one which he had least emphasised. It is the philosophical message of the fractal theory. I'm not so much interested in the mathematical formulation. What interests me is the philosophical aspect. That you basically have to view the world as a very small ball within a huge fractal universe and that the universe is itself a part of an infinite number of other universes.

It's a new view of the world, which is very different from what I learnt at school. That is, when you look at something, to always be aware that you only see a small piece of a larger, bigger whole. And when you analyse it you see how it relates. So we should not just focus on the property of any one individual object, but always see it in perspective and in relation to its larger self."

Benoît: "Yesterday was my birthday. I'm 84 years old. I've had a long and complicated life. It is one that followed very much the pattern of my work. My work has consisted of one major, good idea: *The Fractal Geometry of Nature*. I wrote that book quite a while ago now. And what was the theme of that book? Well, the theme was very, very simple: *Clouds are not Spheres*."

Ivar Giaever: "I like the way he started his book in which he says: 'Why is geometry often described as cold and dry? One reason lies in its inability to describe the shape of a cloud, a mountain, a coastline or a tree. Clouds are not spheres, mountains are not cones, coastlines are not circles and bark is not smooth,

nor does lightning travel in a straight line. What Benoît has done is to develop a new geometry which allows us to describe nature just as we *see* it around us. He has opened our eyes and encouraged us to think in a different way. When I first looked at his book it was the pictures that caught my imagination and that you could actually describe nature with mathematics in such a precise way."

Benoît: "Fractal geometry has provided a language for the description of phenomena which were shapeless before. Fractal geometry is the morphology of the amorphous. It is the tool to grasp phenomena which before were just flowing through our fingers. Let me give two very simple examples. One of them is the variation of prices. The first idea about price variation was that, well, they're predictable. The price of bread is so much then it may go up because of something happening or a bit down. But that's much too simple for financial prices. They never stay constant. The second idea was that prices may in fact be a little bit fuzzy. They can go up and down with the toss of a coin. That was a very good idea, but it didn't go far enough because prices are extraordinarily complicated in their structure. Large events and changes occur, but they are not simple and straightforward at all. The tools to represent them did not exist until very, very recently. Take another example: the shape that has become very famous and to which my name is attached. I'm very proud of it of course: the Mandelbrot set. Looking at it in ordinary pictures, one sees a shape with some circles, with some round forms, but nothing particularly interesting. But as one comes closer and closer and closer, something extraordinary happens. What happens is that one sees more and more detail, more and more complicated designs. The closer you come the denser it gets and also the more beautiful. The Mandelbrot set is totally unlike a circle where if you come closer it just becomes a straight line. It is very uninteresting. Let us suppose that Gauss had invented Gaussian distributions or normal distributions but they had just been lying there because everyone thought they were just some mathematical curiosity. And then someone came along and discovered that there are a tremendous range of natural processes, events and so forth that are best described by normal distributions. And they laid them out one after another and investigated them and then the world woke up and said, "Oh my god, the world is made up of natural distributions." And that's what happened. I took these distribution laws and the notion of fractal dimension which were mathematical curiosities without any significant applications and discovered that half the world is made up of them. I've been concerned with real phenomena — phenomena that nature surrounds us with and not with imaginary shapes which our minds can create and with phenomena that man has in a certain sense contrived, not quite

© Gabriel Lesmoir-Gordon

designed — like the variation of financial prices, like the language of mankind, like the construction of big computers in which the whole design is not quite predictable, does not quite follow from simple rules. The geometry which I learned in school — the geometry with which our whole civilisation is completely filled and permeated — is that of Euclid. It is a geometry of spheres, of triangles, of cones, of circles, of straight lines. But mountains are not cones. Clouds are not spheres. Island boundaries are not circles. Rivers do not flow straight down. Lightning is not straight. The shapes of nature around us are very, very complicated. They are over-whelmingly complicated and they defy description by the tools of Euclid. For this reason, a very large part of the problems which mankind was always faced with — old problems, very old problems, ancient problems — had to be left aside because there was no geometry for them. You could say there was no language to describe them. To do science, you must describe nature. Description is a language. Language can be just words. It can be formulas. It can be shapes. It can be many things. But you do need a language."

Fractal geometry is the language of nature, of the familiar and apparently

random forms like clouds, trees, coastlines, rivers, and lightning.

With fractal geometry, Benoît Mandelbrot has given us a new language which is applicable throughout all the sciences and is one which has a mind opening effect on most people who come across it. This new language is changing our lives and the world of scientific endeavour.

Benoît: "Long ago when I was a young man I had this romantic notion that what I'd like to do before anything else was find some aspects of the world around us which others have somehow left alone, not thought of or not paid attention to. This idea was very much mocked, laughed at and dismissed by my family and friends. When I started implementing it, I started in ways which were quite peripheral. Then I became bolder and found more important topics. And eventually in due course that same quest brought me to some of the most fundamental issues of science."

© Gabriel Lesmoir-Gordon

010

BENOÎT AND NIGEL

Back in the early 1990s, my life took a new and unexpected direction. I was researching topics for possible science TV documentaries and reading Ian Stewart's book *Does God Play Dice?* I came across the chapter he called *The Gingerbread Man*. It was about the Mandelbrot set. Reading that chapter,

I could *see* that here was the perfect subject for a film. The Mandelbrot set is a fractal and fractal geometry is the geometry of nature. So here I had the stunning visual beauty of the set itself and the natural world to point my camera at. It was a science story that I could tell in pictures. It wasn't the Big Bang or gravity or quantum mechanics, it was something you could actually see and which could be filmed directly without recourse to endless and expensive CGI.

There used to be a little shop in Ladbroke Grove, London called *Strange Attractions*. Sadly, it's closed now but at the time it was thriving, selling the products of the new mathematics — fractal geometry and chaos theory. Books, music, computer games, t-shirts, posters, coasters, cups and postcards. Inside, everywhere I looked, I saw the M-set.

Greg Sams, the shop's proprietor, told me that two months before I was there, Fred Clarke, Arthurs C. Clarke's brother, had been driving him back to their apartment in Holland Park through Ladbroke Grove and past Greg's shop. Arthur's sharp eye caught a glimpse of the colourful snapshot of the M-set, swinging in the breeze on a big sign over Greg's shop door. Arthur had Fred reverse the car back to the shop and park. Greg was in.

Arthur had just returned from Riyadh, where he wrote that he had: "*...the privilege of addressing the largest gathering of astronauts and cosmonauts ever assembled at one place (more than fifty, including Apollo 11's Buzz Aldrin and Mike Collins, and the first "space-walker" Alexie Leonov). I decided to expand their horizons by introducing them to something really large!*
So, with astronaut Prince Sultan bin Salman bin Abdul Aziz in the chair, I delivered a lavishly illustrated lecture: The Colours of Infinity — Exploring the Fractal Universe."

Arthur had given Greg Sams permission to publish the speech. When I arrived, Greg had just received the first copies. He gave me a copy of the booklet and Arthur's fax number in Colombo, Sri Lanka and off I went to compose a persuasive letter. I faxed my two-page plea the next morning, went off to work, but forgot to inform my wife, Jenny, that I'd sent it, more or less expecting nothing to come back.

So, when Arthur C. Clarke phoned two hours later, Jenny was stunned and completely unprepared. Only her quick responses saved the situation as she took on board what was happening. She calmed down, took notes and came away with his agree-ment to a two-day shoot at his house in Colombo. And so, with finance raised from family and friends, the filming of *The Colours of Infinity* began.

After we had finished the shooting in Sri Lanka and had returned to the UK, I approached Benoît to ask if I could come to America to interview him. He had the good sense to contact Arthur to ask him if I was the genuine article. Arthur was able to put his mind

Simon Gilbert, John Lamborn, Warner Warnaseri, Paul Sinclair, Arthur C Clarke and Nigel Lesmoir-Gordon

at rest and we set up a time and place to film Benoît.

We arrived at the Mandelbrot family home early in the morning. With my knees shaking I pushed the doorbell. After a few seconds, the door was opened by a smiling and gracious lady, Aliette Mandelbrot. She showed us into the sitting room where we sat and waited. Aliette offered us tea or coffee. The great man came in looking relaxed and comfortable.

While the crew set up the equipment, Benoît and I went to his study to discuss the shape that the interview would take. I had my questions prepared and we went through them. He was gracious and charming and quickly put me at ease. He was essentially happy with what I proposed, making the odd addition and amendment here and there. Benoit was superb, inspiring and lucid.

The next day we flew to Atlanta to interview Michael Barnsley. This shoot went very well. Michael's enthusiasm for the topic was palpable. We completed the filming for this production with an interview with Ian Stewart. David Gilmour of Pink Floyd agreed to write, record the music for the film and I commissioned William Rood to produce the Mandelbrot set animations. With those elements in hand we had all the material that we needed. It took many months to complete the post-production, but when we were done I was truly delighted with the end product. We sent tapes off to the contributors and received warm, positive and encouraging responses from all of them. Benoît was generous with his praise and with his expressions of gratitude. We found a distributor for the film. It went on to sell in over 40 territories worldwide and has been subtitled in three foreign languages. It has been shown on over 80 PBS stations in the USA. Springer published the richly illustrated book based on the film in 2010.

Given the success of *The Colours of Infinity,* I decided to try and make a documentary, which I had been planning, about Benoît's life and work. I was calling it *Clouds Are Not Spheres.* I raised some private funding and won an award from the Alfred P. Sloan foundation to produce the film. I spent two days in Paris in an artist's studio where Benoît told me his life story. This interview formed the backbone of the production. Again, Benoît was eloquent and enthusiastic, never losing his way.

Benoît, Michael Frame and I went on to make the educational DVD production called *Mandelbrot's World of Fractals*, which was commissioned by the National Science Foundation through Yale University. This DVD concludes with Benoît addressing the camera:

"I've spent most of my life unpacking the ideas that became fractal geometry. This has mostly been exciting and enjoyable but it has also been lonely. For years few shared my views. Yet the ghost of the idea of fractals continued to beguile me so I kept on looking through the long, dry years. So, find the thing you love. It doesn't so much matter what it is. Find

the thing you love and throw yourself into it. I found a new geometry; you'll find something else. Whatever you find will be yours. This search is what has carried us from caves to the stars. Where will you take us?"

Apart from being a visionary and a maverick, Benoît was a kind and generous man, always patient and at pains to make his ideas clear and understandable. He was also, in my humble opinion, a genius. The world is unquestionably a far lesser place without him.

At the end of the film *Clouds Are Not Spheres* Benoît closes with this: *"I was extraordinarily fortunate to take the dream of a rebellious and ill-disciplined young man and follow it through. The dream has brought me to many of the most fundamental issues of science. Some I improved, but certainly left very, very wide open and mysterious.*

This had been my hope as a young man. This hope has filled my whole life. In that sense, I feel extremely fortunate."

I have always felt highly privileged to have had the opportunity to work with Benoît and play my part in bringing his groundbreaking work to the attention of a broad and popular audience. Although I had a good grasp of the significance of fractal geometry and brought many years of experience in film production talent to the table I know that I was nonetheless merely a conduit for him. I appreciate now that my films served Benoît well in his efforts to communicate his discoveries and that he was most pleased with what I achieved for him. He always wanted to reach out and communicate warmly across as many platforms as he could and he was a great communicator. When you watch him in my films this will immediately become blindingly obvious. I am deeply grateful to have made three films with Benoît and to write this book. I am eternally indebted to him for opening my eyes to the wonders of the fractal universe and to the beauty of mathematics.

011 AWARDS AND PUBLICATIONS

Benoît Mandelbrot's awards include the 1993 *Wolf Prize for Physics* and the 2003 *Japan Prize for Science and Technology*. Also the 1985 *F. Barnard Medal for Meritorious Service to Science* ("Magna est Veritas") of the U.S. National Academy of Sciences, the 1986 *Franklin Medal for Signal and Eminent Service in Science* of the Franklin Institute of Philadelphia, the 1988 *Charles Proteus Steinmetz Medal* of IEEE, the 1988 (first) *Science for Art Prize* of Moet-Hennessy-Louis Vuitton, the 1989 *Harvey Prize for Science and Technology* of the Technion in Haifa, the 1991 *Nevada Prize*, the 1994 *Honda Prize*, the 1996 *Médaille de Vermeil de la Ville de Paris*, the 1999 *John Scott Award*, the 2000 *Lewis Fry Richardson Award* of the European Geophysical Society, the 2002 *William Procter Prize* of Sigma Xi, the 2004 *Prize of Financial Times/Deutschland*, the 2005 *Orlicz Medal* of Poznan University, the 2005 *Waclaw Sierpinski Prize in Mathematics* of Warsaw University, and the *Casimir Funk Award* of PIASA. He also received a *Distinguished Service Award for Outstanding Achievement* from the California Institute of Technology, and a *Humboldt Preis* from the Alexander von Humboldt Stiftung.

He authored *Les Objets Fractals*, 1975, 1984, 1989 and 1995 (translated into Basque, Brazilian, Bulgarian, Chinese, Czech, Italian, Portuguese, Rumanian, and Spanish) and *The Fractal Geometry of Nature*, 1982 (translated into Chinese, German,

Japanese, Korean, Russian, and Spanish). His *Selecta* began with *Fractals and Scaling in Finance: Discontinuity, Concentration, Risk,* 1997; *Fractales, hasard et finance,* 1997; *Multifractals and 1/f Noise: Wild Self-Affinity in Physics,* 1999; *Gaussian Self-Affinity and Fractals: Globality, The Earth, 1/f Noise and R/S,* 2002; and *Chaos and Fractals: the Mandelbrot Set and Beyond*, 2004. He published his last book *The Fractalist. Memoir of a Scientific Maverick* in 2014.

He co-authored with M. L. Frame *Fractals, Graphics, and Mathematics Education,* 2002 and with R. L. Hudson *The (mis)Behaviour of Markets: a Fractal View of Risk, Ruin, and Reward,* 2004.

A Fractal is a Way of Seeing Infinity.

Benoit Mandelbrot

www.ingramcontent.com/pod-product-compliance
Lightning Source LLC
LaVergne TN
LVHW081930200926
848484LV00009B/2400
9781786344748